FOOTPRINTS

25 COLLABORATIVE WORKS OF GLOBAL ARCHITECTS AND VANKE REGARDING CHINA'S RURAL AND URBAN DEVELOPMENTS

大象无形

中国城乡建设的探索和引领 · 25 个世界建筑师与万科的合作与实践

吴文一 主编

東方出版社

CONTENTS

目录

Preface

序言

在美国读书教书的时候，我曾多次访问一座小城——印第安纳州的哥伦布市，但并不是因为那里有名山大川或者小桥流水。就那座小城的地理风貌而言，它其实是一座再普通不过的美国城市：地势平坦，带有一点点起伏；广阔的农田和大片的森林围绕着只有不到两百年历史的市中心。那几座教堂和几家银行都是美国小城的标配；民居与商业零零散散地混在一起，从深浅不同的颜色和材质各异的房子上可以清楚地看到城市发展的蹒跚足迹。在这座常住居民不到五万人的小城里，有一样东西是与众不同的：这里有百位以上的世界级建筑师和艺术家留下的足迹。他们的作品不但改变了城市的风貌，也让这座小城在世界城市现代史上留下了辉煌的一页。当人们漫步城中，从市政厅、教堂、学校到图书馆，从消防站、企业总部到金融中心，所见的建筑无一不出自名家之手，甚至连城市监狱的设计都能给人们带来别具一格的体验，刷新对生活的感知。而这一切，都起源于小城当地的一家机器制造企业及其业主。对建筑与艺术的热爱以及对生活的诉求和对城市的责任，使他们用自己的企业所得改变了一座普通的城市。他们既让建筑师丰富了自己的职业生涯和人生经历，又让城里的居民和我们这些过路客受惠于大树的荫庇，领略到世界的广袤。

我之所以在本书开篇提到小城，主要有两个目的：一个是分享另一国度的人们对生活与艺术的态度以及为所在城市担负起的责任，另一个则是希望能启发更多的人去关注中国城市发展的现状和未来，期待我们的城市工作者和企业能更真切地了解城市和城市生活，参与其中并尽到更多的责任、发挥更积极的作用。

20 世纪初，英国作家詹姆斯 • 希尔顿写过一本书，名字叫作《消失的地平线》，书中描写的是四名职业背景不同的西方游客来到香格里拉的群山中，他们震惊于那里的神秘与风情，然后在所谓“消失的地平线”中经历了不同的人生故事。随着 21 世纪的到来，世界上又出现了一道新的地平线，中国经济和文化的快速发展给更多有着不同背景的人提供了更广阔的发展空间。在新的地平线前，这些人才冉冉升起，他们的才华和探索精神得到了更多的发挥和认同，城市和乡村因他们而变得更丰富并充满活力，他们也在中国的探索和实践中收获了更多的朋友和一展才华的机会。

在那些远近来客中，我们通过本书给大家介绍的是一批建筑师，一个极具想象力、又热爱生活的群体。这些人中有来自欧洲但从小熟悉岭南大地、现在影响着几代建筑人的领军人物，也有曾在苏州水乡默默耕耘、现在行走于大洋之间的中国学者；有来自东邻友邦建筑领域的老将和新帅，也有偕同女儿坐火车横穿欧亚大陆、带着好奇心来到中国的建筑师妈妈；有为我们留下一片星光后隐身而去的现象级设计师扎哈 • 哈迪德，更有一批来自大江南北、处处求索创新的小伙伴。他们的作品小到河边漂浮“一叶扁舟”，大到横跨几个街区的“水平摩天楼”；有形制简单而穿越时空的地方博物馆，也有肌理复杂、纷繁万象的各类民居……他们的作品代表着我们这个时代前行的脚印，他们执着的探索引导着我们去打开思想的大门。

本书选录的项目都是 2006 — 2018 年间建成或研究的项目。这整整 12 年，在中国文化传统中正好是一个轮回。在这段时间里，这批来自世界各地的建筑师付出了大量心血和劳动，为我们的城市留下了许多长久的印迹。读者可以通过本书跟随建筑师们的脚印一起回望走过的路，领略中国南北大地的广袤及城乡环境的迥异，在每个作品中体会建筑师对不同城市的解读和思考。无论是关注民生问题的公租房还是探索城市形态的综合体，或是锐意革新的学校以及城市社区的改造更新，开卷总会有益。也许你在本书里能找到自己熟悉的城市，或注意到那些遥远而又有几分陌生的地方；也许你能找到一座陌生但又因此而充满吸引力的城市，让你想去熟悉她、接近她；如果你一直在追寻着自己的归属，在本书内外也许能找到一座属于自己的真实的城市。

25 位世界建筑师通过中国城乡的实践案例从不同的角度展示了城乡生活的多样与纷繁。而变化虽然是城市发展的必然趋势，但怎么变却是所有从事城市工作的人和每一个居民都想要知道的问题。通过解析本书的每个项目，人们可以看到建筑师怎样梳理城市脉络、比对城市发展策略、提出解决方案措施。希望这些案例可以给你带来有益的启迪和思考，甚至带你去到各种有趣的城市热点或者比热点更远的远方。

但在去远方之前，我们也需要回到原点，先了解一下这些世界级的建筑师在中国的城乡实践是如何开始的：是谁给了他们前行的机会和能量？

我在本序的开头讲了一个“以社会责任和对建筑的热爱改变城市”的故事，那是发生在 60 年前的大洋彼岸。而今天启动和助力我们这 25 个案例项目，给中国的城乡建设带来具有引导性和综合影响的“理想家”就在我们中间，那就是我们大家都非常熟悉的城市建设者：万科集团。作为一家大型企业，本着“让建筑赞美生命”的理想，通过推行低碳建筑、住宅产业化和绿色物业服务，他们传递的是一种对自然、对社会、对自己与众不同的企业使命感，并以此引领行业，不断用实践印证着他们的追求。通过与国内外的世界级建筑师合作，他们在文化发展、社会责任、大众传播、科技创新等多个领域将城市建设推进到新的高度和广度。我们本次甄选的 25 个实践项目只是万科众多合作项目中的一小部分而已，作为“城市的运营者”他们还有很多项目在大江南北发挥着重要示范作用，服务于大大小小的城市和千千万万的社区。诗人纪伯伦曾说“滴水藏海”，这本合集映射出的也是一种开放的、向上的企业精神。试想若千千万万的企业都以这种精神合作、创业、生活，我们的城市将不仅充满精美有形的建筑作品，我们还将更多地享受到万物生长之大象无形。

吴文一

When I was studying and teaching in the States, I had visited a little town a lot: Columbus in Indiana. This little town attracted me not for any famous scenes. The city is somewhat ordinary in a typical American way: flat terrain with a little rugged hills, vast farmland and forest hug the city center with less than 200 years old of history, a few churches and banks, mixed residential and commercial buildings scattered around with different colors and materialities that showing the history of the urban development through history. This particular town of less than 50,000 regular residents had drawn over a hundred world architects' and with their extraordinary projects, Columbus had become a significant page in world contemporary urban development history. The architecture of city hall, church, school, library, enterprise HQ and financial centers, even the prison design had brought new understanding of city planning, all of which initiated by a local private diesel engine manufacturer and its developer. With their enthusiasm in art and sense of responsibilities in their city, the enterprise had not only changed an ordinary town into a designed-focus city, but also enriched the experiences of design for the participant architects. We, as a pass-by, and the residents in the city, are all benefit from this grand transformation, and indeed, as a consequence, a treasure to the world.

James Hilton, author from England, had written a book named "Lost Horizon" back in the 1930s, in which four western tourists with distinct professional backgrounds had arrived in the mountains of Shangri-La and so overwhelmed by the mysterious sight, they had narrated different life stories in the "lost horizon" landscape. In the 21st century, China had appeared as a new horizon to the world, the rapid development in economy and culture had provided broader opportunities for those had various backgrounds. They had harvested broader collaborative partners while carrying out extraordinary designs in the cities and rural areas in China and been recognized globally for their outstanding talents and adventurous researches, resulted a more dynamic and rich urban-rural-scape.

From this book we have introduced a bunch of extremely creative and enthusiastic architects, who continued to explore the problems of our city, and brought hope with their various schemes. Amongst these architects, one came from Europe,but had known the southern land of China well since his young age, some were local architects from Suzhou and full of experience in different continents. There are architects from Japan and Dutch land, and respected formal architect Zaha Hadid who had left us with her elite works... We could see small projects that floats in the river like a rowing boat, and huge mixed project that crowned as "horizontal skyscrapers", history museums with ultra simplicity in formality, and small residential planning with unbelievably complexity urban textures...These projects represented the footprints of our progressive era, the conceptual safari behind the projects had guided us towards a broader horizon of creative world.

The projects listed in this book were completed or evaluated from 2006 to 2018. The 12-year-transmigration period represented a full samsara cycle in traditional Chinese culture. The architects from the global scene had immersed themselves with substantial works, leaving impresses in the cities of China. We would like to show our readers the paths that we have taken throughout the years from north to south of China, with the narration of the architects, we would appreciate the vast differences between rural and urban parts of the country, hence have a better understanding of how architects interpret the concepts in these areas. We will find projects like public rental buildings in solving livelihood issues, multi-complex project in researching urban morphologies; rejuvenation development for communities and revolutionary ideas for schools...we would always benefit from reading, after all a book that is shut is but a block. You might find familiar cities from near or afar, a city that is strange yet attractive that would worth approaching or visiting in the future. There is always a place for you in this book, a place that you might always be looking for, a place that you would call it home.

The 25 global architects had exhibited a dynamic and complex urban-rural world in their practical cases in China. There is no doubt that we would face changes in the urban development process, each one of us, on the other hand, deserves to know how it would change and what the results might be. Through analyses of each projects in this book, one could realize how the urban texture is rationalized, how developmental strategies are formed and how problems are resolved. We hope that the reading experience would be more than beneficial thoughts in theories, but moreover a further interests for different city places could emerged.

Before heading afar we should also remember where everything had started, how these collaborative works were initiated? How the world architects were motivated and who had been giving out these opportunities?

From the start of the preface I have reflected a story that how an entrepreneur had changed a city with projects full of social responsibilities and enthusiasm for architecture. It happened on the other side of the world about 60 years ago, and now it is happening again among us in China. The idealist that had initiated and assisted the 25 projects is Vanke, a well-known cooperation that had acted as city planner with pioneer concepts and great influences. As a giant entrepreneur, Vanke had a mission to "praise the beauty of life through architecture". They strive for low-carb buildings, housing industrialization and green property services, with leading practical projects to emphasize the enterprise responsibilities for nature, society and themselves. Through constant collaboration with global architects, Vanke had broadened city-planning services to a new extent in terms of cultural development, social responsibility, mass media communication and technological innovation. These 25 projects only represent a small portion in the mass pool of collaborative works of Vanke's. As a "City-Operator",the enterprise had many other classical model-projects in China, serving hundreds of cities and thousands of communities. Poet Gibran once said that "In one drop of water are found all the secrets of all the oceans", the collection in this book also reflected such positive and open-minded corporate spirit. If we could extend such spirit into many other corporations, our cities would be filled with architectural works with aesthetics, entering a world with unlimited imagination with the footprints of the architects.

Wu Wenyi

PART A Solution To The Low-income Group
低收入人群解决方案

01

Tulou Collective Housing

土楼公社

游子的港湾和城堡

建筑设计：都市实践

项目地点：广东南海浔峰洲路万科四季花城

设计时间：2005—2007 年

完成时间：2008 年

用地面积：9,141 ㎡

建筑面积：13,711 ㎡

功能：居住

Archi. design: URBANUS

Location: Guangdong Nanhai

Design process: 2005—2007

Completion time: 2008

Site area: 9,141 m^2

Total area: 13,711 m^2

Program: Residential

土楼是客家民居独有的建筑形式。它是用集合住宅的方式，将居住、贮藏、商店、集市、祭祀、公共娱乐等功能集中于一个建筑体量，具有巨大凝聚力。将土楼作为当前解决低收入住宅问题的方法，不只是形式上的承袭。

土楼和现代宿舍建筑类似，但又具有现代走廊式宿舍所缺少的亲和力，有助于保持低收入社区中的邻里感。将“新土楼”植入当代城市的典型地段，与城市空地、绿地、立交桥、高速公路、社区等典型地段拼贴。这些试验都是在探讨如何用土楼这种建筑类型去消化城市高速发展过程中遗留下来的不便使用的闲置土地。获得这些土地的成本极低，从而使低收入住宅的开发成为可能。土楼外部的封闭性可将周边恶劣的环境予以屏蔽，内部的向心性同时又创造出温馨的小社会。

将传统客家土楼的居住文化与低收入住宅结合在一起，更标志着低收入人群的居住状况开始进入大众的视野。

这项研究的特点是分析角度的全面性和从理论到实践的延续性。对土楼原型进行尺度、空间模式、功能等方面的演绎，然后加入经济、自然等多种城市环境要素，在多种要素的碰撞之中寻找各种可能的平衡，这种全面演绎保证了丰富经验的获得，并为深入的思考提供平台。从调查土楼的现状开始，研究传统客家土楼在现代生活方式下的适应性，将其城市性发掘出来，然后具体深化，进行虚拟设计，论证项目的可行性，最终将研究成果予以推广，这样从理论到实践的连续性研究，是“新土楼”构想的现实性和可操作性的完美结合。

Tulou is a dwelling type unique to the the Hakka people. It is a communal residence between the city and the countryside, integrating living, storage, shopping, religion, and public entertainment into one single building entity.

Traditional units in tulou are evenly laid out along its perimeter, like modern slab-style dormitory buildings, but with greater opportunities for social interaction. By introducing a "new tulou" to modern cities and by carefully experimenting its form and economy, one can transcend the conventional modular dwelling into urban design. Our experiments explored ways to stitch the tulou within the existing urban fabric, which includes green areas, overpasses, expressways, and residual land left over by urbanization. The cost of residual sites is low due to incentives provided by the government; this is an important factor for the development of affordable housing. The close proximity of each tulou building helps insulate the users from the chaos and noise of the outside environment, while creating an intimate and comfortable environment inside.

Integrating the living culture of traditional Hakka tulou buildings with affordable housing is not only an academic issue, but also implies a more important yet realistic social phenomenon. The living conditions of impoverished people is now gaining more public attention.

The research of tulou dwelling is characterized by comprehensive analyses ranging from theoretical hypothesis to practical experimentation. The study examined the size, space patterns, and functions of tulou. The new programs also inject new urban elements to the traditional style, while balancing the tension between these two paradigms. As a consequence of such comprehensive research, the tulou project has accumulated layers of experiences in various aspects. The project provided a platform for an in-depth discussion on feasibilities and possibilites of contextualizing the variable metamorphoses of traditional dwelling modules with an urban reality. It also introduced a series of publications and forums on future hypothetical designs for a "new tulou project". The logic and design process of the tulou program set up a solid foundation and excellent precedent for translating research-based feasibility studies to design realization.

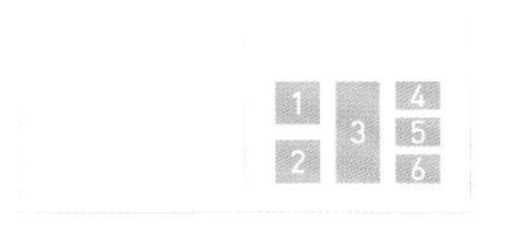

1. 区位图
2. 未来可复制模式想象图
3. 土楼公社轴侧
4. 平面图
5—6. 渲染示意图

1. Site Plan
2. Repetitive Module Future Development Example
3. Axonometric Diagram
4. Typical Plan
5—6. Renders of Tulou Collective Housing

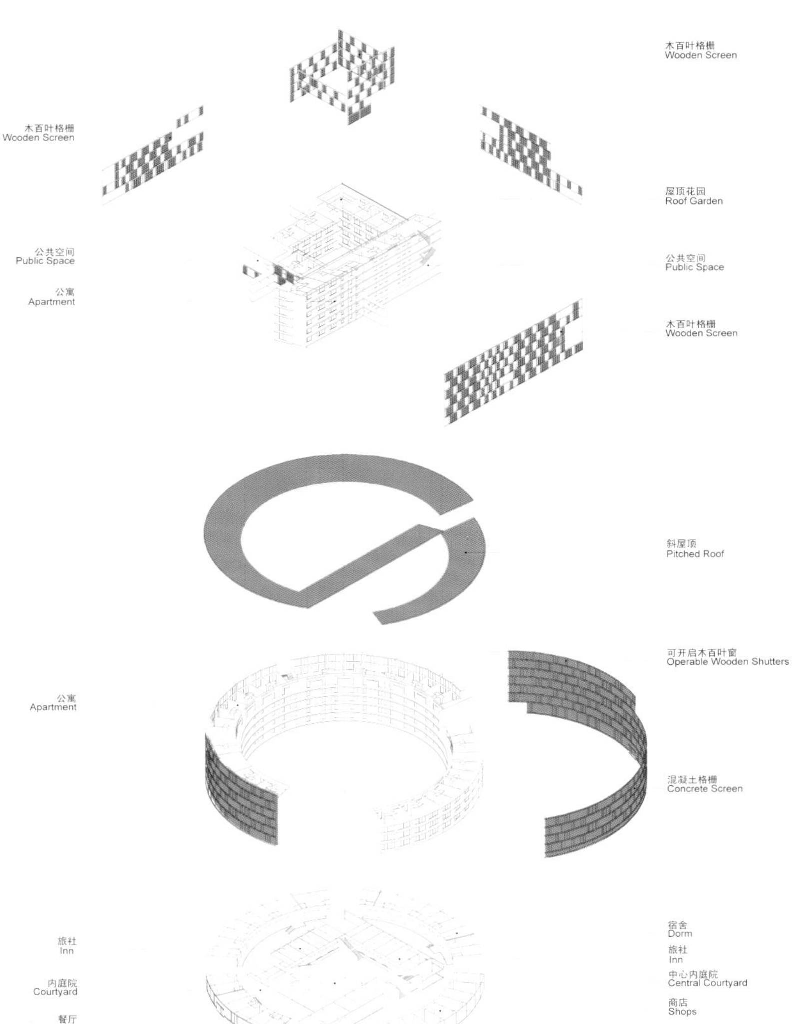

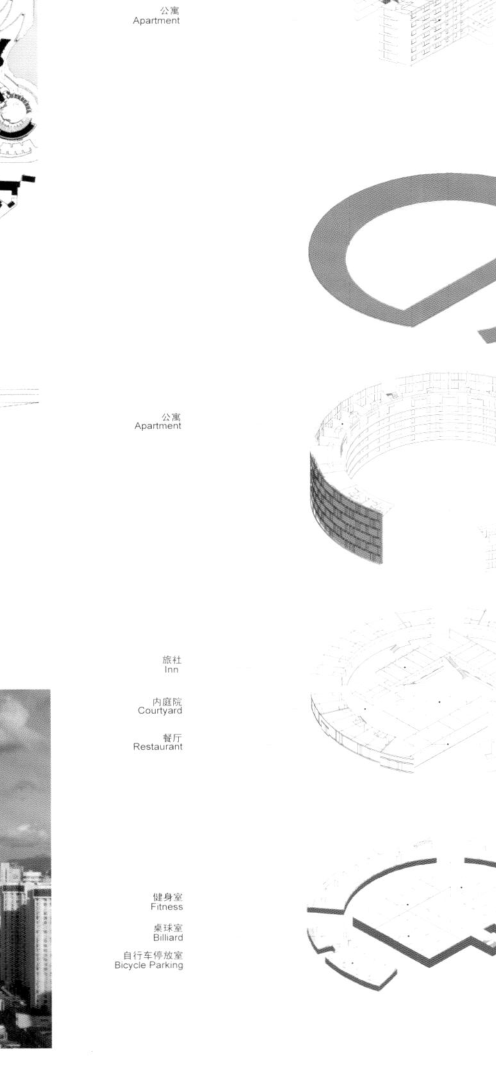

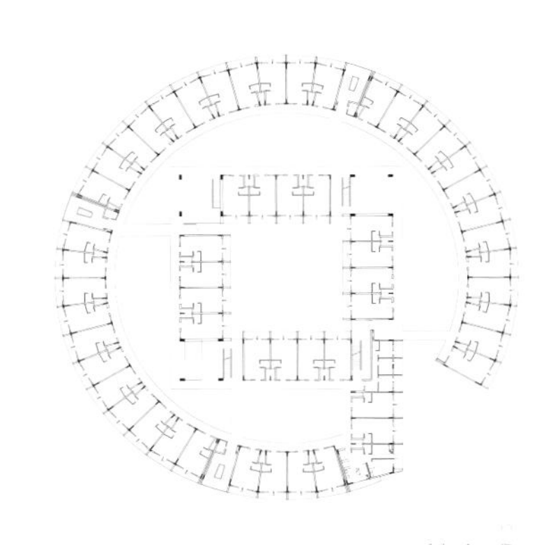

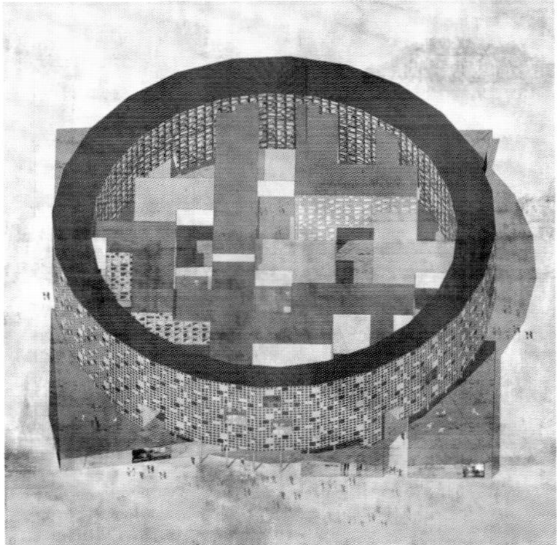

土楼公舍 Tulou Collective Housing

“尊重平凡的城市建筑物和城市人，是城市可持续发展的基本前提。”

关于建筑师

1999 年，刘晓都留美归来便一头扎到深圳，做深南大道改造的街道景观美化，为他和孟岩、王辉刚在纽约成立的都市实践探路。项目虽小，但他们为城中弱势群体设计了公共活动空间，这一创意据说引起了当地政府的注意，随后都市实践便不断接手公共建筑和城市设计，在与社会发展接轨的同时坚持介入当下，包括向客户和城市决策者传递他们的理念。为解决深圳城中村的问题，2005 年底万科与都市实践开始讨论低收入住宅开发的可能性，王石提出了土楼的概念，想以传统客家民居为原型，做符合三保人员经济条件和社交需求的廉租房。然而，在深圳“第五园”做试点的消息一经公布就遭到小区业主出于阶层意识的反对，使选址最终转移到广州。但正如刘晓都所说，低收入者也有“居住的权利”和“生活的尊严”。作为改善外来务工人员和低收入者居住状况的公共住宅实验，土楼公社不只是自成天地的复合建筑，也是城市更新的优质样板。

ABOUT THE ARCHITECT

In 1999, after Liu Xiaodu returned from his stay in the United States, he absorbed himself with a project in Shenzhen to beautify the street landscape of Shennan Avenue and explore new methods to expand the scope of Urbanus which had been established in New York with Meng Yan and Wang Huigang. The project was small in scope, but they designed public space for vulnerable groups in the city and as a result this idea attracted the attention of the local government. Subsequently, Urbanus have continued to take over public buildings and urban design, while keeping in line with social development by insisting on intervening in the present and engendering their reasoning with customers and urban decision makers. In order to solve the problem of villages in Shenzhen, at the end of 2005, Vanke and Urbanus began to discuss the possibility of developing affordable housing. Wang Shi proposed the concept of "Tulou", which involved a desire to construct low-rent housing that meets the economic conditions and social needs of individuals reliant on the three-insurance, applying the traditional Hakka residence as the prototype. However, as soon as news of the pilot project of Zone 5 in Shenzhen was announced, the proprietors of the community objected out of class consciousness and the site was ultimately transferred to Guangzhou. However, as Liu Xiaodu stated, low-income people also have the "right to live" and deserve "dignity of life". As a public housing experiment with the goal of improve the living conditions of migrant workers and low-income people, the Tulou Collective Housing is not only a self-built complex architecture, but also an ideal model for urban renewal.

1. 都市实践 — 南山婚礼堂

1. URBANUS Nanshan Wedding Chaple

刘晓都
Liu Xiaodu

孟岩
Meng Yan

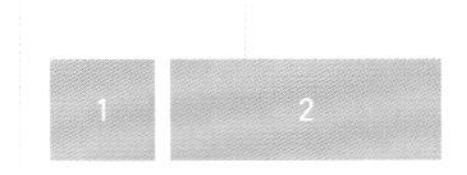

1. 土楼公社中庭内景
2. 远眺土楼公社

1. Interior View of the Courtyard of Tulou Collective Housing
2. Tulou Collective Housing from Afar

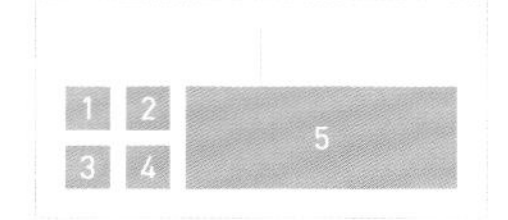

1—4. 土楼公社内部实景
5. 永定客家土楼

1—4. Tulou Collective Housing Interior Views
5. Traditional Yongding Hakka Tulou

星艺装饰
XINGYI DECORATION
电话：020-38818321
GOLD APPLE ALUMINIUM

02

Wan Cun Rejuvenation Plan

万村复苏
让城市美好共生

建筑设计：坊城建筑设计
项目地点：地处福田区，北靠南园路，
南临滨河大道，与香港新界隔河相望
建筑面积：约 126,700 m²
楼栋：114 栋
项目类型：城中村改造

Archi: FCHA
Location: Futian Distrct, North Bound To Nanyuan Road,
South Bound To Binhai Road,
On The Riverbank Opposite To Hongkong New Territory
Total Area: Approx. 126,700M²
Building Numbers: 114
Program: Urban Village Regeneration

万村改造，更新城中村生态

“万村复苏”计划——在保障村民收益的前提下，以集约化整体租赁的形式，对村民自建物业进行专业化改造，开创性地提出“综合整治 + 内容运营”的新思路，通过硬件升级（增建基础设施、改造景观环境、建设休闲广场、新增居住服务休闲设施、美化建筑立面形象等）和城中村文化内容生态重构（为城中村导入居住、商业、教育、文化等配套内容）对城中村进行改造整治。改造后的城中村将作为万科城市配套、服务业务的落地平台，计划纳入长租公寓、物业管理、社区商业、社区营地等业务板块。

“WAN CUN PLAN”, A PLAN TO UPDATE THE URBAN VILLAGE ECOLOGY

The Wan Cun Plan is a new strategy to upgrade the hardware of the village (including additional infrastructure system, reconstruction in landscape and public plaza, additional service and leisure system and upgrade in building facade etc) and the software of the cultural ecology (such as providing commercial, educational and cultural supports to the villagers). The plan combines overall rejuvenation with unified management by the developer, ensuring the income of the current residents. After the rejuvenation plan, the village would transform into a service support platform of VANKE's city system, and long-term rental apartment, property management, community spaces will also be introduced.

1. 爆炸图

1. Explosive View

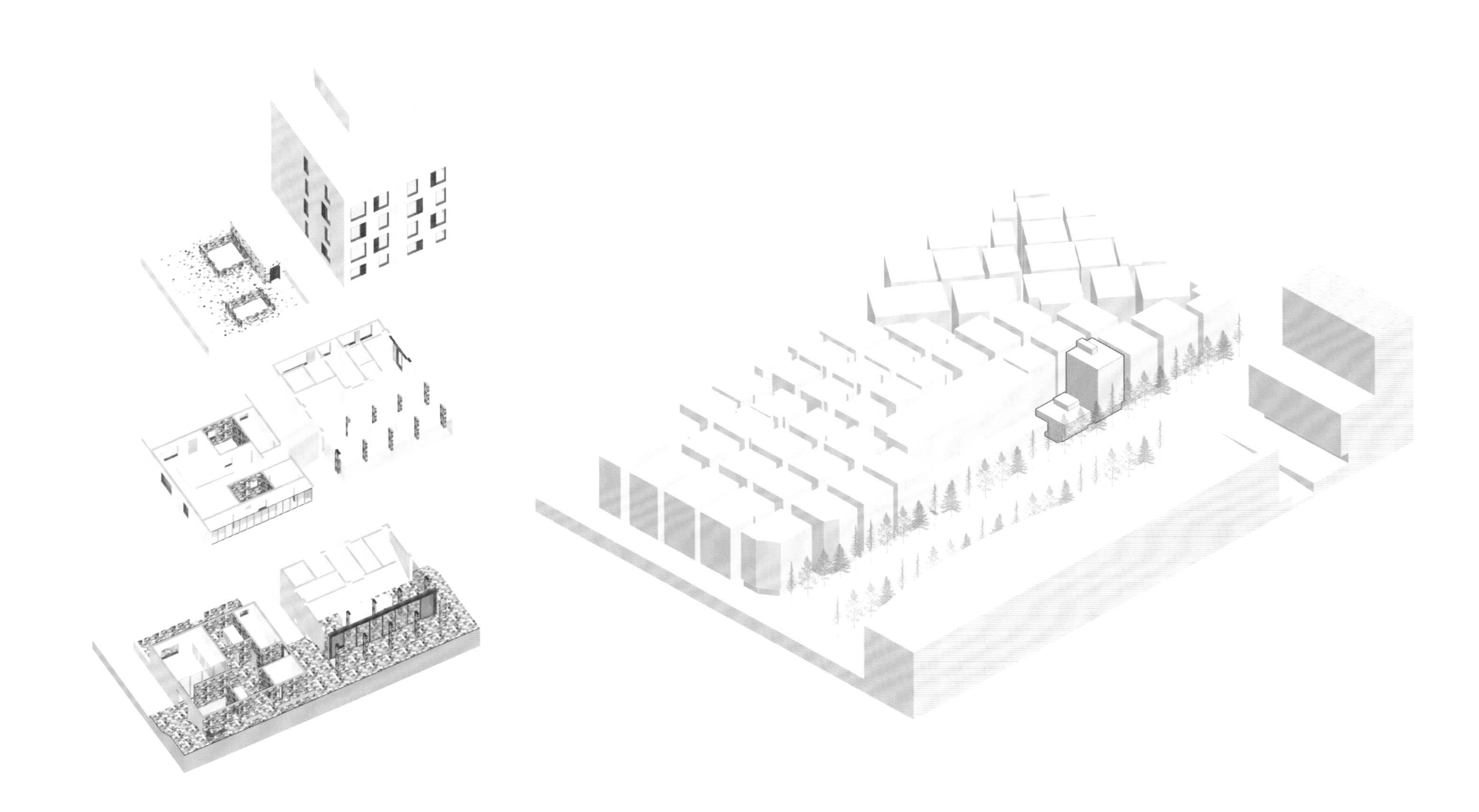

“大部分的项目都要在约束条件非常高的情况下去寻找有趣的点，不管有趣与否，这些项目往往牵涉很多人的生活质量还有城市景观，我们不能掉以轻心。”

关于建筑师

坊（Fang）：街坊，是构建中国古代城市的基本元素，也是具有传统文化意味的城市公共空间。

社区（Community Blocks），城市中最重要的模块单元，是决定城市空间质量及尺度的重要元素，是由街道围塑出的空间结构。

城（CHeng）：城郭，是市民集中居住、生活的载体，意指公司将从更加广阔的城市范围研究定位我们的建筑设计，思考城市发展，并研究新的城市模型及城市未来。

建筑（Architecture）：组成我们城市的最基本单元，是组成街坊、城市的重要元素，类型丰富，尺度多样，涵盖所有与人类生活密切关联的建筑物。

坊城设计（FCHA）：坊城设计传达了我们对待城市、街坊、建筑的设计理念，既要从更广阔的城市分析来切入街坊及建筑设计，同时也要在建筑设计中体现街坊的邻里关系及生活，也表达了我们以建筑类型学这样一种研究方法对中国城市化进程的思考。

ABOUT THE ARCHITECT

Fang: Streets, they are the basic elements that constituted the ancient China urban fabric, they were the public spaces with sense of traditional culture in the city.

Community Blocks: the vital modules in the city, they form the scale and quality of urban spaces, and construct their own spatial structures with the embrace of surrounding streets.

CHeng (City): a platform where people live, the company would focus the design and research base on a broader concept based on a city scale, emphasizing new urban models and future prospectives.

Architecture: the basic physical module that forms the city, architecture is an important element that constitutes the streets and the urban environment, the types and scales vary base on different needs.

FCHA: a company that translates our concepts of Fang, CHeng and Architecture. We analyze architecture design with a sense of urban planning, and care for neighborhood and living style during the design process, producing research methods with architecture typology in China's urbanization process.

陈泽涛
Chen Zetao

1. 五角环 —— 泰丰幼儿园
2. 总规平面图

1. Pentagon Loop — Taifeng Kindergarden
2. General Layout

城市中心福田玉田村——城中村复兴领跑者

玉田村地处福田区，北靠南园路，南临滨河大道，与香港新界隔河相望，地段繁华，位置优越，成为无数深漂人首选落脚点。项目内建筑面积约 126700 平方米，约计楼宇 114 栋。

由于历史遗留原因，玉田村建筑密度奇高，地面空间拥挤不堪，采光较差，存在严重消防隐患；空间呈现组团式特质，各组团空间缺少有机联系，交流困难；村内交通管理封闭，入口不明显；建筑形式多为楼间距离极近的“握手楼”，城市空间匮乏；教育资源紧缺……如今的玉田已为租户提供约 1500 套长租公寓；约 5000 平方米公共屋顶花园和约 1000 平方米首层户外活动空间也按时开放，大大提升了城中村活力。作为核心区城市更新标杆项目，目前总签约楼栋共 40 栋，均在施工更新中。玉田村的改造是万科“城乡建设与生活服务商”这一价值理念的成功实践。

YUTIAN VILLAGE — PIONEER IN URBAN VILLAGE REJUVENATION

Yutian Village locates in Futian District, its elite location includes Bin He Road to the south, Nan Yuan Road to the north and only a river across to Hongkong New Territories district. The overall planning area is 126,700 sqm with approximately 114 buildings involved.

Yutian Village has a high-density building ratio and over-crowded ground space due to historical issues. The existing urban fabric has bad light condition and fire hazards. The spaces were not connected, they are random clusters like isolated islands. The traffic is problematic, the distance between buildings is one can hardly tell the entrance of the village and the community is lack of management, urban spaces and educational resources are also weak in the area. Now after the rejuvenation development, Yutian Village has 1500 long-term rental apartments for local needs, there are approximately 5000 square meters of public roof gardens and 1000 square meters of ground floor outdoor activity spaces for a more dynamic life style. As a model-project of CBD rejuvenation, there are 40 buildings with renovation contracts under construction. The Yutian Village project had become one of the successful practice of the concept of “servicing the life and construction of city and village” by Vanke.

1—2. 原来的城中村
3—4. 改造前后的玉田村

1—2. Previous Urban Village
3—4. Yutian Village Before and After Renovation

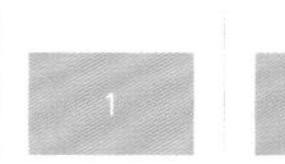

1—2. 改造后的玉田村

1—2. After Transformation

PART B Truly China
骨子里的中国

03

Shenzhen Zone 5

深圳第五园

乡土中国的典雅

建筑设计：王戈工作室

项目地点：广东省深圳市龙岗区布吉镇坂雪岗南区

设计时间：2003—2005 年

完成时间：2005 年

建筑总面积：100,000 ㎡

功能：合院别墅、联排别墅、多层住宅

Arch. design: BOA Studio

Location: South Ban Xue GANG, Buji Town,
Longgang District, Shenzhen, Guangdong

Design process: 2003—2005

Completion time: 2005

Total area: 100,000m^2

Program: Courtyard Houses,Town Houses,Multi-story Residential Building

万科第五园整个社区的规划由中央景观带分隔而成的两个边界清晰的“村落”所组成，一条简洁的半环路将两个“村落”串连。每个“村落”都由三种产品即庭院House、叠院 House 以及合院阳房所构成。各“村”内部都由深幽的街巷或步行小路以及大小不同的院落组合而成，宜人的尺度构成了富有人情味的邻里空间。位于社区紧邻城市干道的商业街和社区图书馆与住宅区之间以池塘相隔，小桥相连，互为景观。其内部空间也特别强调了各种开敞、半开敞、下沉的院落和连廊组合，形成丰富而使人流连的“村口”场所。

在住宅单体上，庭院别墅的“前庭后院中天井”以及通过组合形成的“六合院”和“四合院”，叠院 House 的“立体”小院（院落＋露台），合院阳房的围合所形成的“大院”，种种院落形式无不着力体现中国传统民居当中那种“内向”型的空间，提供了一方自得其乐的小天地。项目设计中重新思考了空间上的开合和趣味，单独的院落和房屋以村落的形态聚集，有其自身的空间秩序。由城—村—过巷—进院—入房，极大地延长了对中式居住空间的体验过程，这是单一建筑的传统空间所无法达到的，对这一脉络的梳理使设计慢慢找到了线索。

从风格的定位上来看，它既不是讨论过几十年的“大屋顶”，也不应是来路不明无可考据的怪诞形象；在电影《卧虎藏龙》的启示下，第五园的建筑风格谨慎地选择了中式建筑的代表——徽派民居和苏州园林。它的整体形象最接近于现代建筑风格，有很大的发挥余地。在外部形象上，着力塑造墙体的形态，又不使其对于通风采光造成影响。一色的青灰色金属瓦楞屋面，平直而无起翘，配合墙脚灰色质感涂料，墙头的简洁压顶则可以游走在传统与现代之间。

作为潮湿炎热地区的中式建筑，该项目吸收了富有广东地区特色的竹筒屋和冷巷的传统做法，通过天井、廊架、挑檐、高墙、花窗、孔洞、缝隙、窄巷等，试图给阳光一把梳子，给微风一个过道，使房屋在梳理阳光的同时呼吸微风，让居住者能充分享受到一片阴凉，在提高了住宅的舒适度的同时有效地降低了能耗。

第五园，试图用白话文写就传统，采用现代材料、现代技术和现代手法，创造一种崭新的现代生活模式，但不失传统的韵味。国内很多地方流行制造所谓“欧陆风情”的虚假的都市景观，第五园则逆风而起，以文化的自信，表现“乡土中国”中饱含的典雅和精致。整个社区形成类似于传统村落形态的具有人情味的丰富的邻里空间，能够为社区居民之间的交往提供舒适的公共空间。

第五园也因此被评为第一届中国建筑传媒奖居住建筑特别奖入围作品。

1. 老房子平面手绘图
2. 老房子手绘图

1. Old House Plan
2. Old House Sketch

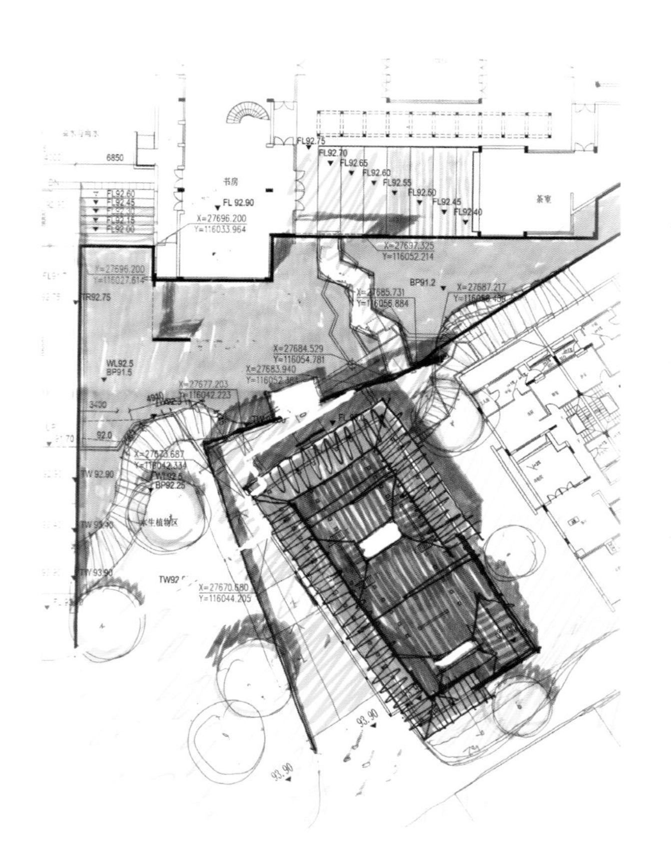

The entire community of Vanke Zone 5 has been planned to consist of two distinct "villages" separated by a landscaped belt that straddles the center of the two villages. Each "village" is made up of three kinds of products, which includes the courtyard house, superimposed yard house and integrated sun room. Each "village" is composed of deeply serene alleys and walk ways in addition to courtyards of differing sizes. The comfortable size and lay out of the area constitutes a humane neighborhood. The commercial street and community library located close to the main street of the community are separated from the residential area by ponds, and connected by small bridges, providing for exceptional views from both sides. The interior space also emphasizes various open, semi open, and sunken courtyards and gallery combinations. The result is the formation of a "village entrance" full of interest which compels people to delay their departure.

For the individual resident building unit, the courtyard villa is formed by a front yard, a backyard and an additional small yard in the center. In this way the unit forms a so-called "six-sided courtyard" and also "four-sided courtyard". The townhouse has a three-dimensional yard formed by the yard and terrace, whilst the "courtyard" is formed by the enclosure of integrated sunroom. All kinds of courtyard forms reflect the kind of "introverted" space in traditional Chinese dwellings, and provide a small space for people to enjoy themselves. In the design of this project, Wang Ge thought about the openess and interest of the space, and in addition to the individual courtyards and houses which gather in the form of villages that display their own spatial order. The experience of the Chinese living space has been greatly extended from the city to village, and from the alley to courtyard, before finally entering the home. This is impossible to achieve in the traditional space of a single building, however we slowly found clues by combing through this thread of design.

From the perspective of style selection, neither the "big rooftops" that have been debated for decades, nor unidentified grotesque image of unknown origin were utilized. Rather, inspired by the movie Crouching tiger and Hidden Dragon, the architectural style of the village was carefully chosen to be representative of Chinese architecture— namely, elements of the Huizhou residence and Suzhou Gardens. The overall image of these kinds of buildings is most close in nature to the modern architectural style, which provides ample space for experimentation. Combined with the gray textured coating on the wall, the minimal top of the wall features a style with both tradition and modern elements.

As a Chinese-style building in humid and hot areas, the project drew lessons from traditional Guangdong architecture such as thick bamboo tube houses and deserted alleyways. By utilizing the patios, galleries, eaves, high walls, windows, holes, gaps and narrow lanes, it was possible to attempt to cool the effects of the sun and allow the building to breath. The ultimate aim was to allow the dweller to fully enjoy the shade, and improve the comfort of the house while effectively reducing energy consumption.

Shenzhen Vanke Zone 5 has attempted to use vernacular language to convey tradition, by using modern materials, modern technology and modern techniques in order to create a brand-new style of modern life without losing any of the traditional charm. The whole country is constructing the so-called "continental feeling" of a false urban landscape. In contrast, Zone 5 moves against the wind, with a sense of cultural self-confidence, and a mission to show the elegant and refined qualities of "native China".

The Zone 5 had been nomivated for the First China Architectural Media Awards Special Award for Residential Buildings.

1. 商业区域修改草图
2. 立面图

1. Commercial Zone Sketch
2. Elevation

"从'注重细节'到'忘掉细节'，如何让建筑体现出'神韵'才是一个建筑最为关键的。"

关于建筑师

ABOUT THE ARCHITECT

王戈 1995 年从天津大学硕士毕业后进入北京市建筑设计研究院工作，2007 年凭借有"骨子里的中国"之称的深圳万科第五园获得"亚洲建协荣誉奖"。他说"每样东西都有一个基本点"，第五园的基本点就是通过研究徽州老房子得来的一种"神韵"或"气质"，让运用现代材料、技术和手法建造的中式建筑散发出乡土民居的质朴气息和诗意韵味，并体现出"村、墙、院、素、冷、幽"的中国特色。同时他也提出，他的作品并非都是纯正中式，而是表达了精神上的"中国风"，做东方人喜欢的静谧院落，有平缓惬意的生活气息而不是矫揉造作的张扬激越。在他看来，好的建筑就四个字："恰如其分"——该高调的时候高调，该低调的时候低调，与周围环境相适应，绝不彰显特立独行。

After his graduation from Tianjin University for his master studies, Wang Ge worked in BIAD in 1995. His project "Vanke Zone 5, China to the heart" had won "ARCASIA awards for Architecture" in 2007. Wang Ge believes that for each element there is a fundamental core, in his renowned Zone 5 concept the core is the elegance spirit from the traditional houses in Hui Zhou. Such simplicity and poetic vibe is achieved with modern materials and building methods to further embody the Chinese concept of "Village Wall Courtyard, Plain Calm Serene". On the other hand, he explains that his projects are not "purely-Chinese", but rather to achieve a sense oriental feeling of calmness and layback lifestyle in the courtyards. He points out that a good architecture should be "Appropriate", which means a harmonic existence with the surrounding environment, being iconic only when needed.

王戈
Wang Ge

1. 甘肃和政古动物化石遗迹馆
2—4. 模型

1. Hezheng Museum of Ancient Animal Fossils,Gansu
2—4. Models

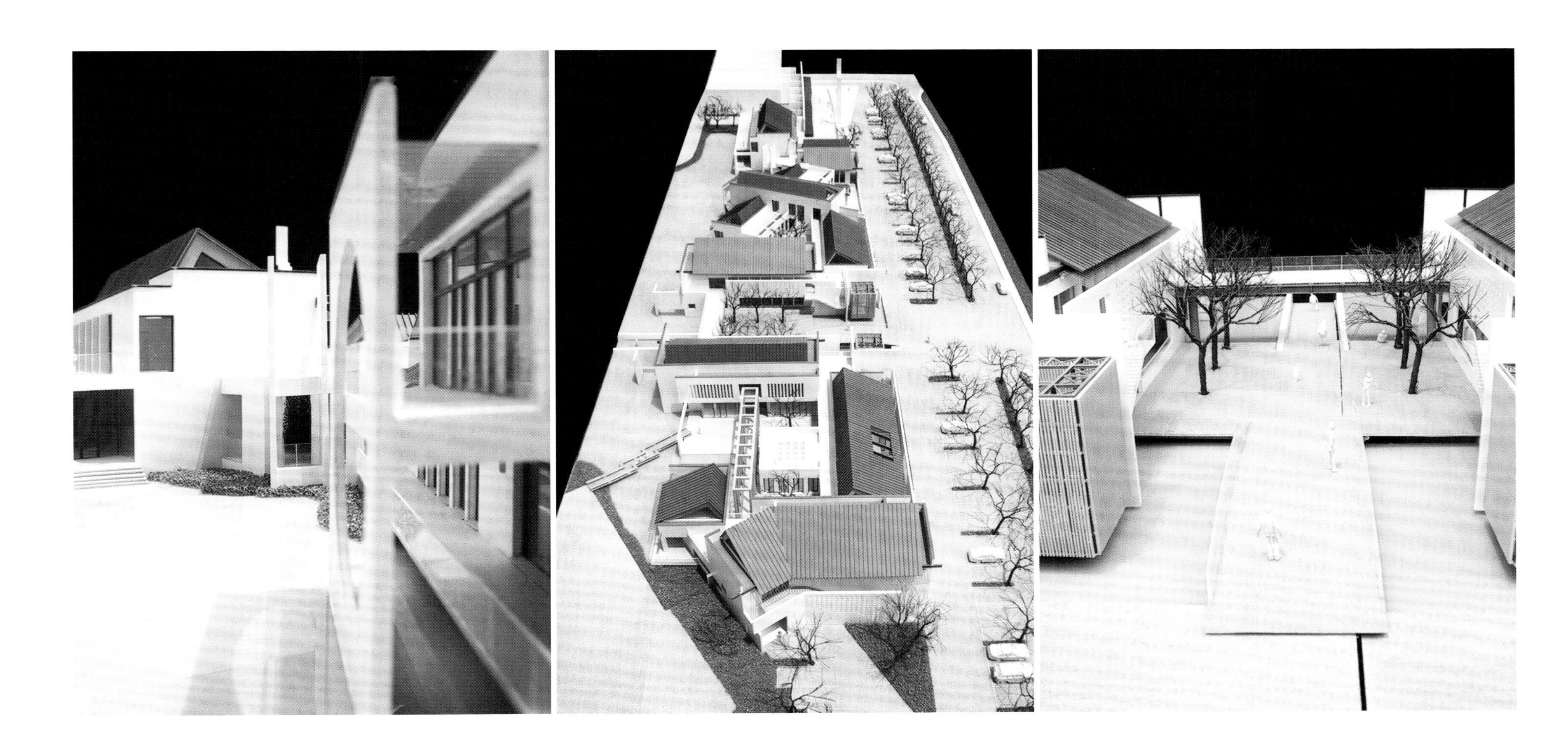

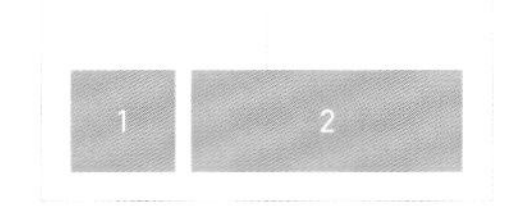

1. 室内图
2. 水岸风景

1. Interior View
2. Waterfront View

04

"An" Housing Project

"本岸" 集合住宅院落

中国院落之本，当代住宅的新岸

建筑设计：标准营造
项目地点：苏州工业园区金鸡湖东岸
设计时间：2006—2007 年
完成时间：2007—2008 年
会所面积：3,700 ㎡
功能：健身会所、社区服务

Arch. design: Standard Architecture
Location: East Shore of Jinji Lake, Suzhou Industrial Park.
Design process: 2006—2007
Completion time: 2007—2008
Club area: 3,700m²
Program: Gym, Community Service

“本岸”是中国集合住宅的创新类型，它不同于一般的联排式住宅，而是把院落重新结合在住宅之中，丰富了中国当代的居住方式。新院落住宅基地紧邻两条河流，基地周围原散落着一些傍水而居的自然村落，为典型的江南水乡景观。规划系统以传统村落鱼骨状自然生长结构成片段，片段构成组团，组团围合院落，而院落又与铺砌着石板的街巷相连，营造了现代的院落空间。

“岸”会所位于社区的北界，既为社区的门户所在，也充当社区私宅部分与城市道路间的屏障。会所自南施街沿琼姬路东向绵延五百余米。车行所见，是隔着一带渐宽水面的树影中时断时续的白墙，视线下沿是银光乍闪的三角锥草坡。白墙延续着“标准营造”对连续折叠的兴趣，也延伸着他们对墙面虚实趣味的追求。人行道上的两排 31 个三角锥草坡沿中线错落，间或以镜面不锈钢板覆面，映照着沿街的城市速度，别出新意；放低视点，“片山多致，寸石生情”，水面及植树白砂草坡带则恍惚成了带硬边的米家云山。

枕流并排的 13 个带天窗的盒子 / 院子，塑造了两个层次的围护关系：一个半包围的大院子中一座贴着北界的楼——1.5 m 厚的墙限定基地的东界和南界，东西向伸展的建筑主体则以 13 道互为向背关系的剪力墙划分。再次一级的空间划分则是南北向的 4 道分隔。这如许多的互成角度、轻重有别的纵横分割，成就了房间、院落、中庭、水体、天桥、步石穿插的空间复调。东界厚墙北段内藏直跑楼梯，上行过天桥可穿入建筑主体的二楼，北段下沿则是 2.2 m 的会所主入口；南界厚墙与主体建筑间是竹林一片，墙体西端安置着设备用房出口和又一段直跑楼梯，下沿通过 4 个 2.2 m 的开口并一带雉堞状凹凸收边的水面，与其南侧的住宅区隔水相连。

“岸”会所的盒子进退、院落穿插显然远较前两例复杂，而墙面肌理却是最平整的白色粉刷。东西向面街的墙体中，窗玻璃靠墙体内沿，以保证沿街立面的墙体厚度效果；南北向墙体中则设有多处 600 mm 高的“墙底窗”，玻璃居中，将波光水影带入室内，乃至起舞染上室内墙面。“岸”会所也灵活悠然地重演了“石头院”自“空院”开场的空间体验序列，从东界厚墙 2.2 m 高的门洞步入，面对的是香樟、榉树荫影婆娑的高墙，向右侧身，院子向水面开敞，天桥凌水而过。

会所由一面长 500 余米的连续墙体转折而成，其间围合成十三个相对独立的院落。微妙的斜线切成的路线引导人在其中享受“曲径通幽”的体验。每个开间都为 不规则的瓶口状，开间与开间相互咬合形成内宽外窄的内向空间；横向的桥把开间切成不同功能的空间。室内、院落、中庭、水体和桥，这些元素相互交错，使得苏州趣味再次呈现。

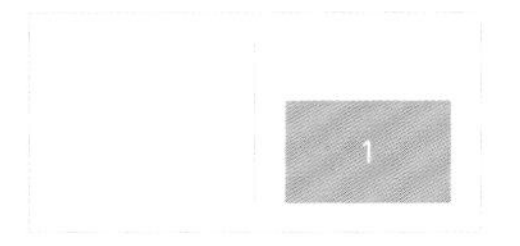

1. 手绘设计过程，2006

1. Sketch Design in 2006

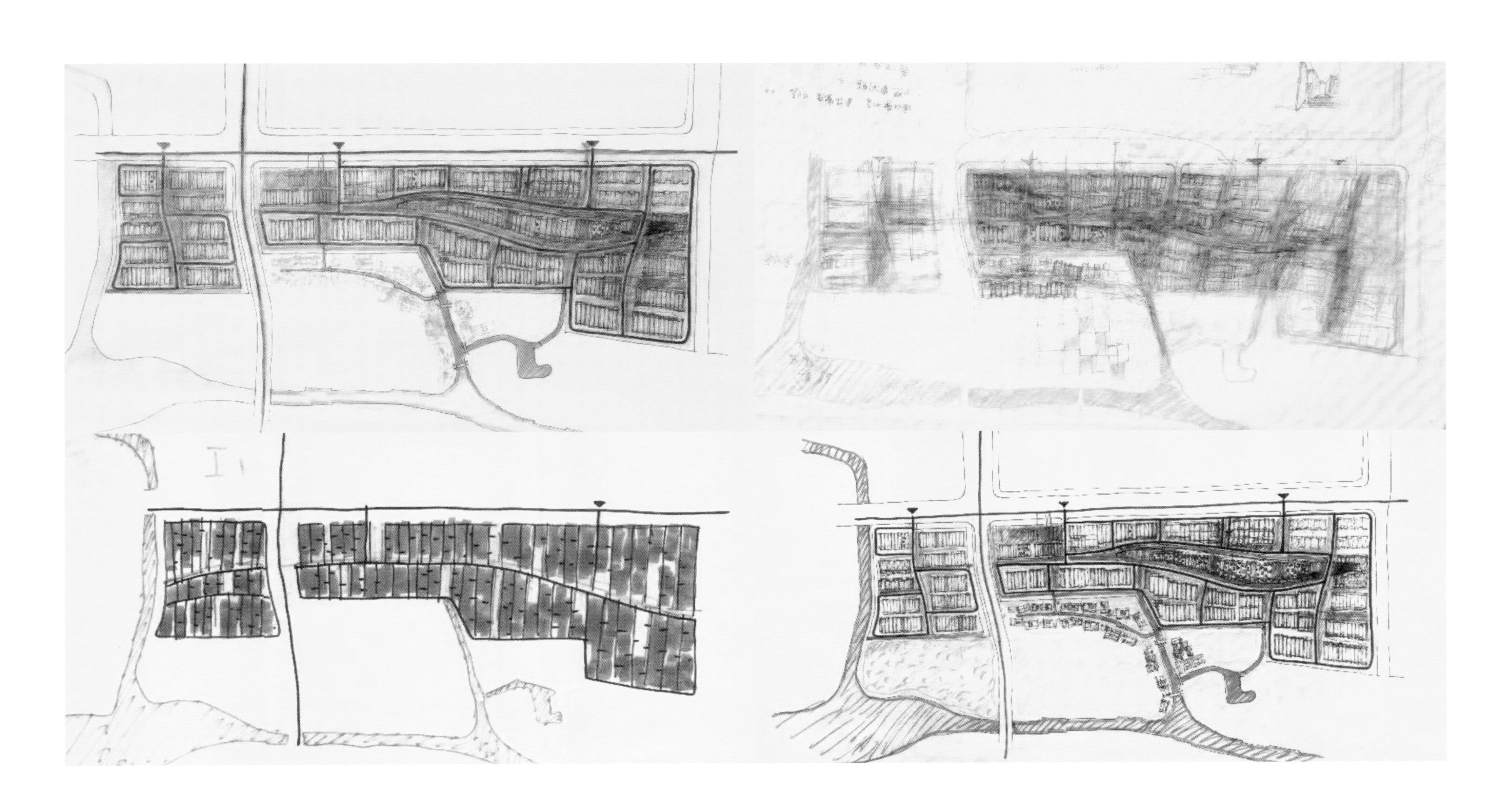

"An" is an innovative type of collective housing in China. It is different from common row housing, as it integrates the courtyard into the housing structure, so as to enrich the Chinese contemporary living style. The new courtyard residential base is adjacent to two rivers. Around the base is a scattering of natural villages that are located near to the water. It is a typical water village landscape often seen south of the Yangtze River. The planning system is composed of fragments of traditional fishbone-like natural growth structures in the villages. The fragments form groups which surround the courtyards. The paved streets and lanes of the courtyards are connected to create a modern courtyard space.

The "An" clubhouse is located by the northern boundary of the community. It is not only the gateway of the community, but also a barrier between the private houses and the urban road. The clubhouse stretches over 500 meters from south to south along Hainan East Road. The view from the passing cars is an intermittent white wall set as the backdrop between the shadows of the trees that gradually widens, and below the line of sight are glistening triangular cone-shaped grass slopes. The white wall continues to generate interest from the continuous overlapping of the "standard architecture", yet also extends the pursuit of interest towards the wall with regards to reality and illusion. With regards to the two rows of 31 triangular cone-shaped grass slopes along the middle line on the sidewalk, the design features mirrors or stainless steel plate surfaces which reflect the speed of the city as it passes along the street by adopting an original approach. This approach is encompassed in lowering the point of view. It is as if the surface of the water and trees planted in the white sand and grass slopes are in a trance with Mijiayun Mountain.

Thirteen skylighted boxes/courtyards are positioned side by side, which creates two levels of enclosure: a half-enclosed courtyard with a building close to the northern boundary —a 1.5m-thick wall that defines the eastern and southern boundaries of the base; and an east-west extension of the main building that is divided by thirteen shear walls that are mutually supporting. The second level of spatial division is found in the 4 separations between the north and the south. This features a myriad of mutual angles, varied weights, and offers different vertical and horizontal divisions. The result is the realization of the room, courtyard, atrium, water feature, bridge, and stepping stones, which are interspersed with a spatial polyphony. There is a straight flight of stairs in the northern section of the thick wall of the east boundary, which provides access to the bridge so as to access the second floor of the main building. The lower part of the northern section is the 2.2 meter main entrance of the club. In the South section, there is a bamboo forest positioned between the thick wall and the main building. The west side of the body is equipped with a room for equipment, exiting and features another straight flight of stairs. The lower edge passes through four 2.2 meter openings as well as a corner of the surrounding water feature area. It is connected to the residential area on the south side.

The arrangement of boxes and the interposition of the courtyards in "An" clubhouse are obviously more complicated than some of the earlier projects done by Standardarchitect, while the texture of the wall is of smooth white paint. With regards to the east-west street wall, the window glass runs along the body of the wall to ensure the overall effect of the wall's thickness along the street facade; as for the North-South wall, there are many "footer windows" of 600 mm in height that are featured with glass-centered design. The waves of light and shadows stream into the interior, and even dance as different hues across the interior wall. "An" clubhouse has an elegant and casual experience of spatial sequence. Upon entering from the 2.2 meter high door of the east wall, one is met the fragrant scent of camphor and the shadows of the beech trees across the high walls. To the right, the courtyard opens to the water and the bridge overflows the sparkling surface of the pond.

The club is formed by a continuous wall that is more than 500 meters long, which encloses thirteen relatively independent courtyards. The path, which has been subtly carved by oblique lines offers people a "winding path" experience. Each opening is irregular, and bottle-shaped, which provides for an interaction between the openings. The result is the formation of a space that features a narrow exterior, whilst providing for a vast, spacious interior. The transverse bridge separates the opening into different functional spaces, and each element, including the interior, courtyard, atrium, water feature and bridge. All of the elements are intertwined, once again bringing a sense of playfulness to Suzhou.

1. 本岸院落模型

1. An Courtyard Model

"除了尺度之外，我们还应该一直思考人性，即我们怎么对待在城市生活里面的个人。"

关于建筑师

在西藏林芝做项目时，张轲带着自己的团队一砖一瓦地造房子，一盯就是个把月，对材料工艺亲力亲为，随时按现场需要调整设计。他为人低调和蔼，很会生活，做事却高调执着，精益求精。这让他与多位年轻设计师在1999 年纽约创办、2001 年迁至北京的标准营造成为业界标杆，将一丝不苟的匠人精神贯彻得极为彻底。无论在西藏娘欧码头，还是在北京茶儿胡同，张轲都擅长因地制宜，让建筑融入景观，将可持续和人情化做到淋漓尽致。标准营造自觉不受各类潮流的影响，而是仔细探究项目地点的周围环境和人文内涵，使每件作品都植根于历史和社会状况，保持清醒和克制，展现思想和论辩。这种对话式美学使苏州"岸"会所建立了一种内向型的公共空间，与当地园林的建筑传统相契合，营造了"曲径通幽"的微系统体验。如今在哈佛执教的张轲这学期所开的课程就围绕苏州园林和诗意栖居展开，曲线救国式地活跃在建设中国的"战地前线"。

ABOUT THE ARCHITECT

During the project in Tibet, Linzhi, Zhang Ke personally took his team to build houses brick by brick. The time passed rapidly, and soon the work had progressed over a couple of months. He personally handled the materials and technology, and adjusted the design according to the needs of the site throughout each stage of the project. His personality is low-key, kind, possessing the ability to live and work whilst maintaining a high profile and constantly endeavoring to do better. This resulted in him and many young designers founding an agency in New York in 1999 and subsequently moving to Beijing in 2001, where StandardArchitect has become a benchmark within the industry, and the meticulous spirit of craftsmanship is featured through vigorous implementation. Whether the Namchabawa Pier in Tibet or the Chacate Hutong in Beijing, Zhang Ke excels in adapting to local conditions, integrating architecture into the landscape, in order to achieve both sustainability and humanization. StandardArchitect remains unaffected by various trends, instead carefully exploring the surrounding environment and humanistic connotations of the project site, so that each work is rooted in historical and social conditions, maintaining a sense of sobriety and restraint in order to display ideas and arguments. This concept of dialogical aesthetics allows the "An" clubhouse in Suzhou to establish an introverted public space, which is in line with the architectural tradition of the local gardens and creates a "winding path" micro-system experience. Today, Zhang Ke, who teaches at Harvard, has commenced his course this semester around gardening and poetic dwelling in Suzhou, and is active in constructing China's "battlefront" by his own means.

张轲
Zhang Ke

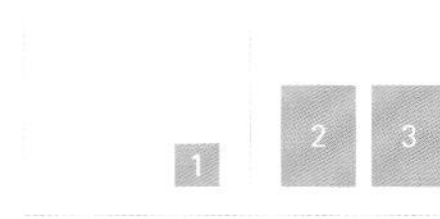

1. 西藏南迦巴瓦访客中心
2. 十二年前，2006 年 10 月
3. 十二年后，2018 年 10 月

1. Tibet Namchabawa Visitor Centre
2. 12 years ago, October 2006
3. 12 years later, October 2018

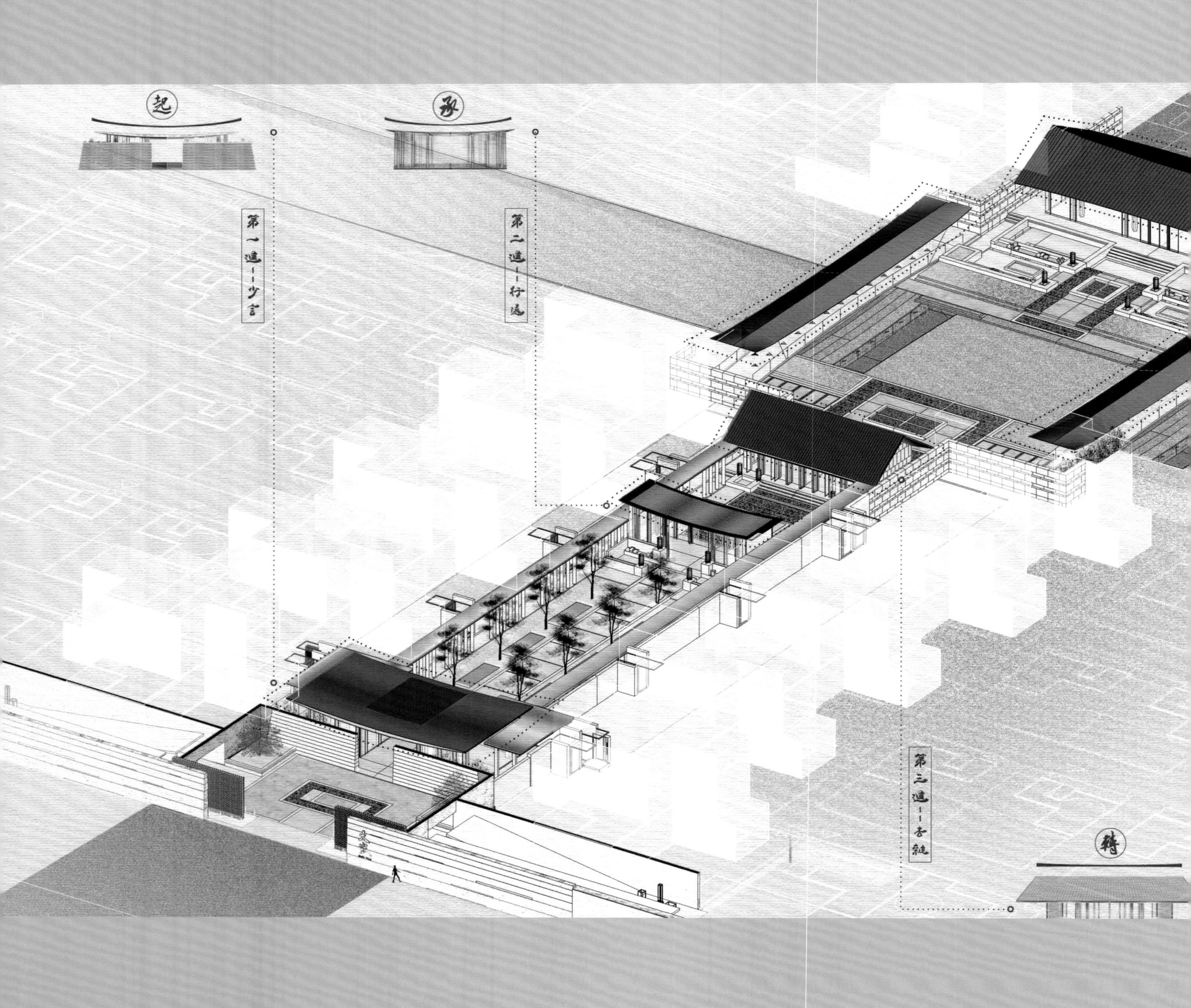
起
承
第一進——少言
第二進——行遠
第三進
转

05

DAJIA Villa

大家

大境为序，仁人成家

建筑设计：上海日清建筑设计有限公司
项目地点：苏州工业园区
设计时间：2016 年 12 月—2017 年 08 月
完成时间：2017 年 12 月
功能：住宅

Archi. design: Lacime Architectural Design
Location: Suzhou Industrial Park.
Design process: 2016.12—2017.08
Completion time: 2017.12
Program: Residential

相比于着眼独栋单体的设计和推敲，中国传统建筑更加关注对建筑群组的控制，赋予群组中各个单体的体量、比例、朝向、抱合以宗法礼制层面的意义，显示出对建筑之间的节奏控制和自然人文的重视。在“大家”中轴文化区的设计中，将中国传统建筑的这一特点与建筑形象和空间节奏进行了充分的结合，分布于轴线上各个建筑单体中，起承转合间，以“廊”贯穿院落，以“檐”起势造境，层层递进。

设计中主要采用现代化的材料和工艺，旨在突破传统建筑因结构和材料所限制的形式束缚，以独立于时代的工艺与来源于传统的空间产生独特的化学反应。对传统建筑不加甄别，在设计中，以铝镁锰板、超白玻璃、香槟金铝板取代了传统的小青瓦和红砖，以高强度大跨度的钢结构和与之相应的构造节点代替了抬梁穿斗的木构架。现代材料不同于传统的质感和肌理，以及现代结构和施工工艺对形式和空间的突破，使“大家”中轴文化区满足了与当下要求相一致的建筑形象与功能空间。

在“大家”社区有一东西向市政河流横穿而过，割断了地块间南北向的联系，且该市政河流并不在社区的用地红线之内。设计的难点在于如何能跨越红线，使整个社区统一为一个有机整体，并产生城市层面的积极意义。因此，贯穿整个社区的人文轴线应运而生，与景观轴线中正相交，而两个轴线交汇之处，便成为整个设计的“题眼”所在。最终，两座全钢结构的轻盈桥体，其形象抽象于中国古代的虹桥，形成了两道优美的弧线跨河而过，双桥基础在南北地块红线之内，而桥体跃于红线之外，完成了对社区南北地块的连接。这座双桥可被视为“催化剂”，它的作用机制可以被解释为：凭借桥梁物理性质的连接，使得中轴线南北两侧的建筑和双桥一道，组成一个完整的院落，使其能作为一个向外开放的核心，成为一个面向公众的，可以用来交流、休憩、活动的城市客厅，以促进城市街区的良性发展。特别是基地处于以居住用地性质为主的城市街区中，缺乏公共交流场所，双桥的介入更是能产生一加一大于二的效果，使得一处消极的、内向型的、私密的宅间用地变成积极、外放、公众的城市客厅，改善和促进城市街区功能的自我调整。

“大家”项目继入围 WAF 世界建筑节，获得 ICONIC Awards 之后也顺利地被提名 WAN 世界建筑新闻奖 & 德国设计大奖；就连大型公建都很难企及的世界结构大奖 the Structural Awards 也一并入围，起到了很好的表率作用。

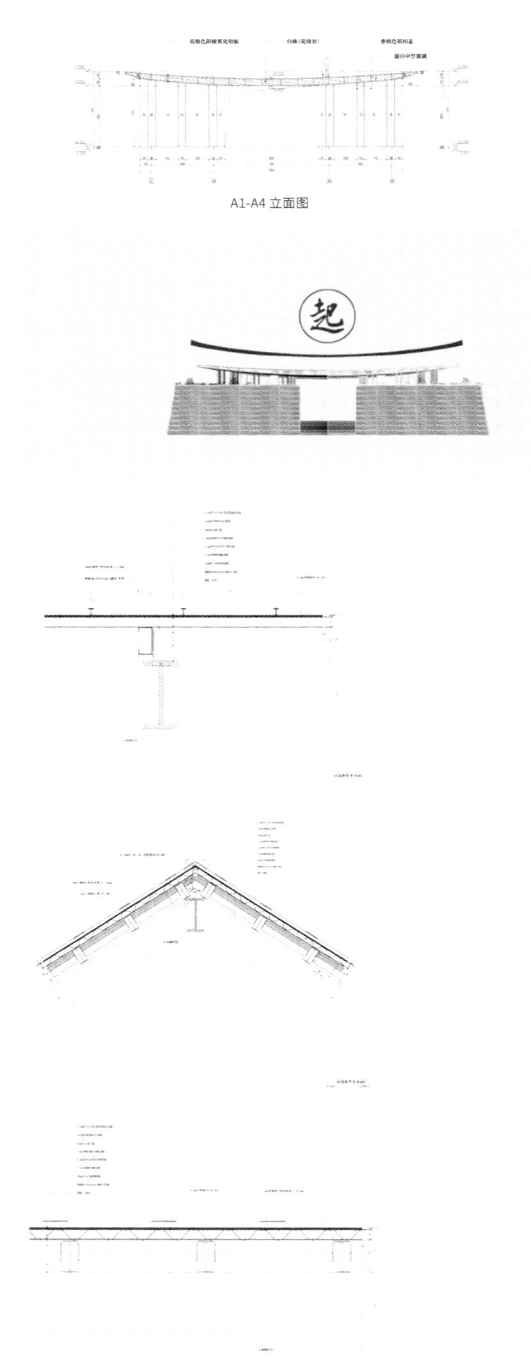

A1-A4 立面图

坡屋顶 T 型锁边方法

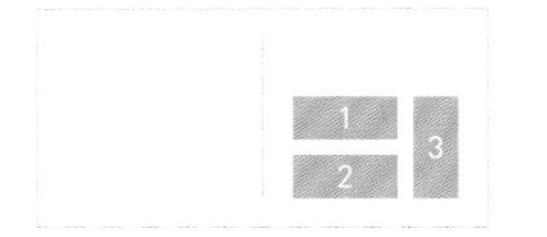

1. "起" "承" "转" "合"的变化
2. 构造细节
3. 文化中轴线长卷

1. Opening, Developing, Changing and Concluding in Form Modification
2. Construction Details
3. Cultural Axis Long Scroll

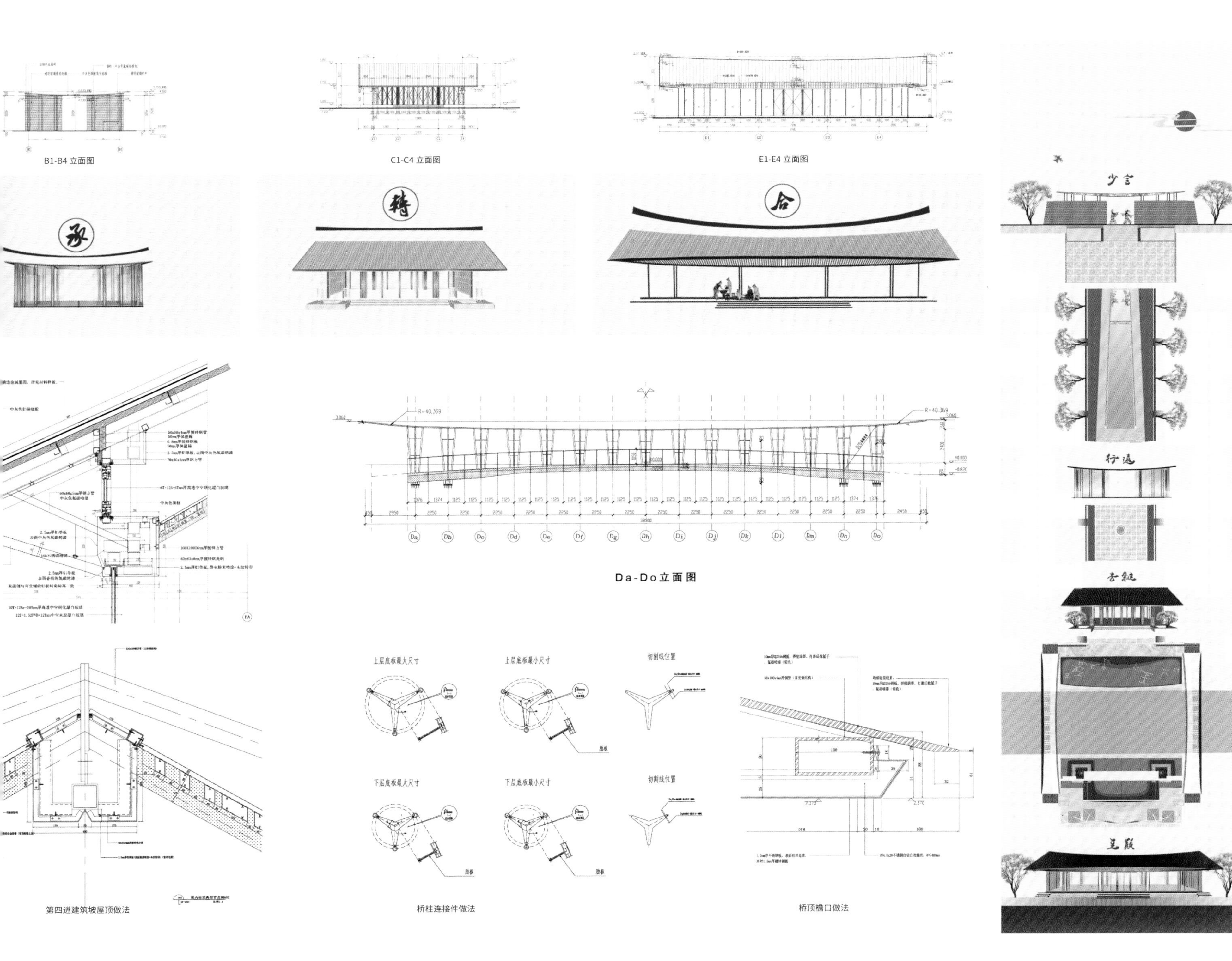

Traditional Chinese architecture would emphasize on the rhythm and "Li" (the etiquette system) lies behind the overall cluster of buildings rather than focus on the architectural design on one particular entity. In the cluster system, the "Li" element implies to each building in regards to volume, scale, orientation and enclosure, indicating the importance of the "groove" and culture between buildings. DAJIA overall cultural axis planning has a spacial rhythm that combines the traditional Chinese architecture feature and the modern forms, the verandas are the linkages between the courtyards and the eaves provide spacial layers in both visual and physical experiences.

The modern material and building methods had explored possibilities with less restrains, breaking through the traditional form of architecture which is retrained due to construction limits. There is chemical reaction between the independent craftsmanship and traditional spacial layout, replacing traditional materials of Chinese style tiles and red brick with Aluminum Magnesium Manganese plates, Super Clear Glass and Champion-gold colored aluminum plates. High strength steel structure and details are used instead of the wooden bracket sets, the texture and structure from modern materials and construction methods had broken through the boundaries of the old form, fulfilling the needs of the current lifestyle and functional space requirements.

Dajia Community has a municipal river crosses, cutting off the north-south connection between the blocks, and the municipal river is not within the red line of the community. The difficulty of design lies in how to cross the red line and unify the whole community into an organic unit. Consequently, the human-axie of the whole community in north-south direction emerges as the times require, centrally intersecting with the axis of the landscape, and where the two axes intersects thereby becoming the subject focus of the whole design. In the end, the two light-weight bridges of all steel structure which abstracted from ancient China style rainbow-shaped bridge form two beautiful arcs across the river. The double-bridge foundation is within the red line of the north-south lock, while the bridge body jumps over the red line, completing the connection of northern and southern blocks of the community. The double bridge may be regarded as the catalyst. The buildings on the northern and southern sides of central axis and the double bridge form a complete courtyard by virtue of the connection of physical properties of bridge, making that it can serve as a core opening outside and become a public living room which can be used for communication , leisure and activities to promote healthy-sound development of urban blocks.

After being nominated in the WAF (World Architecture Festival) and won with ICONIC Awards, DAJIA Project, developed by Suzhou Vanke, is further nominated in the WAN Awards and German Design Award. It is also nominated in the Structural Awards, which is an elite award highly unreachable even for some large public construction projects. DAJIA becomes a pinoneer model in international design competitions.

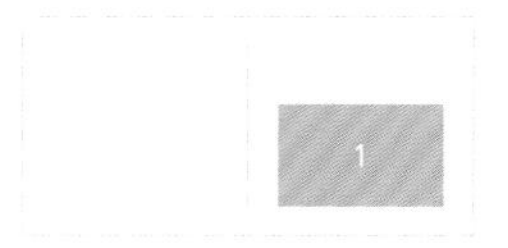

1. "大家"中庭照片

1. DAJIA Courtyard Photo

“建筑的本意就是情感的形态和空间，通过一系列图像去打动心灵的最深处。以极简的笔法，在线条与材质的架构下，成就一个生活的艺术场景，一个精神的空间场所。”

关于建筑师

宋照青从小就通过父母了解建筑行业，在西安和清华学习后于 1995 年出国加入日建设计，从绘图员做起，到 1998 年参与上海新天地改造的设计和施工，逐步成长为注重实践经验的建筑师，并创立了以“删繁就简，溯本清源”为宗旨的本土公司日清设计。日清真正接手的第一个项目是苏州都市花园，而到建设“万科 • 大家”园林小区时，宋照青已在苏州和国内其他城市完成诸多获奖工程。在日建期间他就注意到国外建筑师敬业务实，很少谈论流派，而是“从现场出发，以最简单的思路去解决住居、交流、环境的问题”，所以宋照青觉得建筑本身“没有过多玄妙的东西”，只是“情感的形态和空间”。这种形态空间可以像深深打动过他的敦煌博物馆内从地下伸展出来的青砖墙体一样打动人心，因为对他而言，建筑就应该这样结合传统和现代，既要成为文化载体，又要符合时代特征。

ABOUT THE ARCHITECT

Song Zhaoqing learned about the construction industry from his parents during his formative years. After studying in Xi'an and Tsinghua, he went abroad to join NIKKEN SEKKEI LTD in 1995. He started as a draftsman and participated in the design and construction of the Shanghai Xintiandi rennovations in 1998. He gradually developed into an architect who paid attention to practical experience. He also established a local company called LACIME Design with the purpose of "Simplify complicated material". The first project that Nissan Design took over was the Suzhou Urban Garden. During the period the "Vanke DAJIA" garden villa project was being constructed, Song Zhaoqing had already completed many award-winning projects both in Suzhou and other parts of the country. When he was involved in projects in Japan, he noticed that foreign architects were respectful and practical, and seldom talked about genres. Instead, they, "start from the scene, with the aim of solving the problems of residence, communication and the environment by applying the simplest way of thinking." Thus, Song Zhaoqing maintains the philosophy that architecture itself "does not possess many mysterious things", but simply "emotional form and space". This form of space can be as touching as the green brick wall that extends from the ground in his Dunhuang Museum. For him, architecture should be combined with tradition and modernity in this way, and should be both a conveyance of culture as well as possessing the characteristics of the period.

宋照青
Song Zhaoqing

1. 旭辉集团大楼
2. “大家”社区夜景
3. “大家”中庭冬景

1. CIFI Building
2. DAJIA Night View
3. DAJIA at Winter

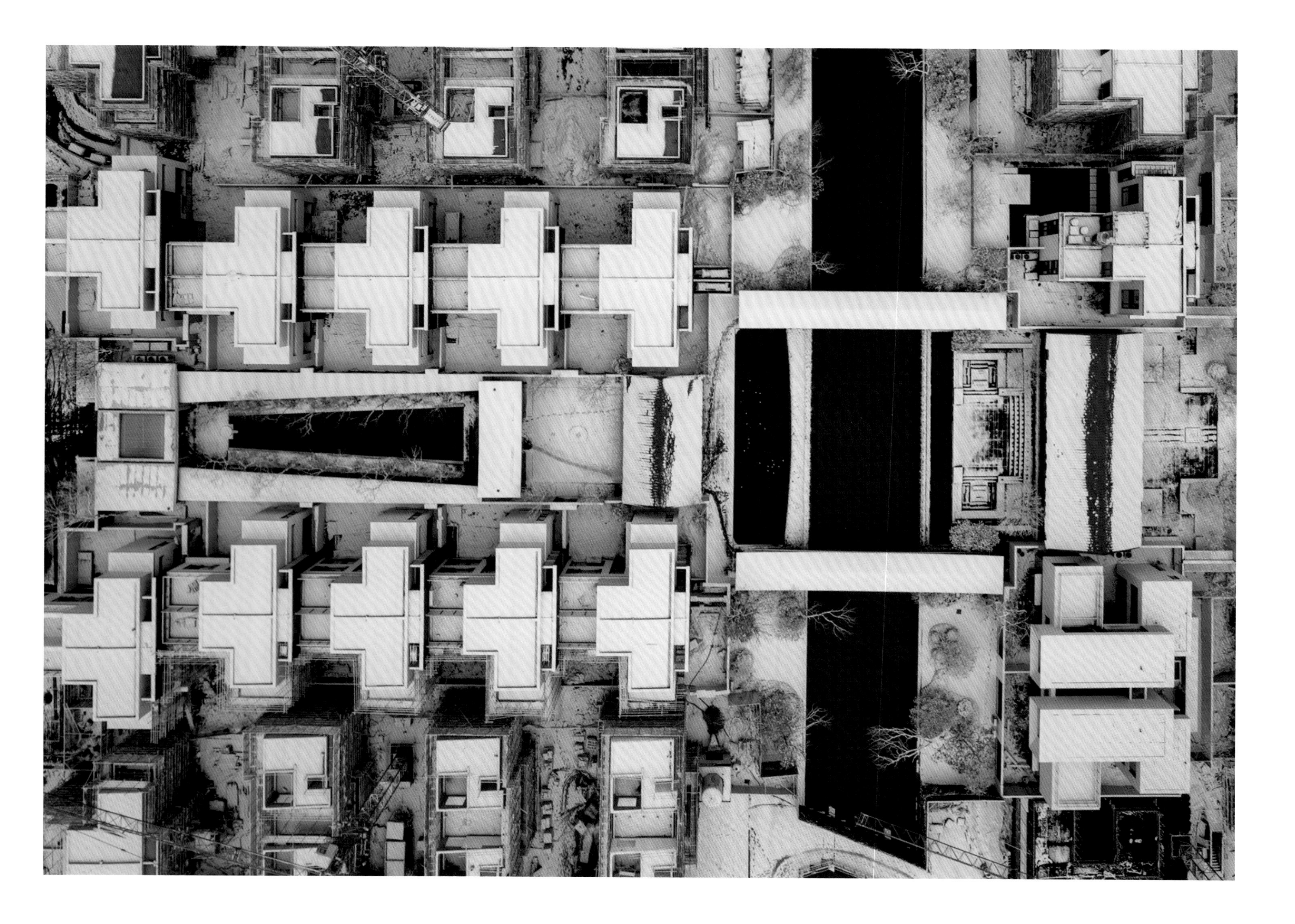

1. 总规划图平面图
2—4. “大家”社区

1. Masterplan
2—4. DAJIA Villa and Community

06 DAXIANG

大象山舍

以世界语言转译东方造园精髓

建筑设计：SCDA 建筑师事务所
项目地点：苏州高新区
设计时间：2016 年 12 月
建筑面积：285,539 m²
功能：住宅

Archi. design: SCDA
Location: Suzhou High-tech District
Design process: 2016.12
Total area:285,539m²
Program:Residential

大象山舍，万科大字系全新作品。

项目位于苏州高新区新乐园版块凤凰峰路与罗家湾路交汇处，被大阳山森林公园、新苏州乐园环绕，自然生态和景观资源丰富，周围配套资源齐备。

汲取全球高端住宅的设计经验，新加坡 SCDA 建筑师事务所在“大象山舍”的方案中，对建筑、自然和人居三者之间的关系提出了独到的见解，将高端品质、人居美学以及功能实用性有机融合，塑造出国际一线的高端住宅产品。产品内容涵盖叠拼、高层等多种类型，布局呈散落式排布，并以光线决定建筑位置及朝向。

在建筑形体的设计上，SCDA 采用了极简的线面结合，以彰显建筑的外形视觉张力。立面采用全玻璃幕墙，建筑立面光洁而轻盈，并引入三面景观，让社区景观与自然和谐共振，并以世界语言转译东方造园精髓。

DAXIANG is the latest project revealed by Vanke as part of their "DA" series.

The location of DAXIANG is at the intersection of Feng Huang Feng Road and Luo Jia Wan Road at Suzhou National Hi-Tech District (SND), the site is surrounded by Dayangshan Forest Park and the new Suzhou Amusement Land with great natural landscape resource and completed infrastructure system.

With SCDA's renowned experience in elite residential design, the DAXIANG project is reflecting the design team's unique perspective with regard to the relations between architecture, nature and user. The scheme infuses the high-end quality design with lifestyle aesthetics and functionality, resulting an international-recognized product. A series of different types of residents, including stacked villas and high-rises, are planned in an organic scattering manner, with which their orientations are designed base on light.

The form of the architecture is simple with clean and sharp lines, powerful in visual impact. The facade is fully glazed to ensure views on three sides while maintaining a light-weighted feel, the community is harmonically merged into the nature, revealing an eastern-garden essence with a global perspective.

1

1. 大象山舍鸟瞰图

1. DAXIANG Birdeye View

“设计应从具有代表性和民间的风格中解放出来，将建筑设计的体量和立面建立在体积、光线和表面这样的原型元素之上。”

关于建筑师

ABOUT THE ARCHITECT

曾仕乾在马来西亚槟城的龙山堂邱公祠长大，从小就常随从事房地产开发的父亲去建筑工地。他先后在华盛顿和耶鲁就读，硕士毕业后实习了三年才到新加坡发展，于 1995 年创立 SCDA，既做设计师，又做开发商。他喜欢古典主义的申克尔，也喜欢现代主义的里特维尔德和新艺术派，但作为世界文化遗产一部分的邱公祠里那古老的院落，以及天井等增加平衡感的“负空间”，至今仍影响着曾仕乾的设计，使他爱用天窗和庭院，让自然光影填充空间。他的设计平静优雅有层次感，形成高品质的“新热带风格”。从新加坡国家设计中心，到自带泳池的纽约空中别墅 Soori High Line，再到体现园林美学的苏州万科大象山舍，曾仕乾的作品遍布全球，获奖颇丰，也使 SCDA 成为尤其以高端豪宅见长的世界知名建筑事务所。

Soo K. Chan grew up in Khoo Kongsi in Leong San Tong, Malaysia. When he was a child he followed his father, who worked as a developer, to different construction sites. After graduated from Washington and Yale, Chan spent three years as intern before he returned to Singapore. In 1995 he founded SCDA, worked both as chief designer and developer. As a fan of the classical Schinkel, Chan also obsessed with modernism in Rietveld's design and followed the Art Nouveau influences. However, his background in ancient courtyard in the world heritage site Khoo Kongsi had rooted in Chan's design. The balancing feel of the negative spaces created by courtyards, the skylights and shadows caused by natural lights... Chan's design elements abundant with elegance and layers with a distinct high quality of "Neo Tropical" style. From Singapore National Design Center, to Soori High Line in New York, and the most-current garden style Vanke Daxiang project, Soo K. Chan's design had great achievement across the world and SCDA had become an iconic international design studio with specialties in elite housing.

曾仕乾
Soo K. Chan

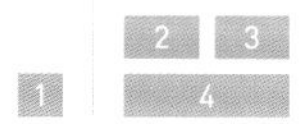

1. 纽约 Soori High Line
2. 大象山舍渲染图
3. 大象山舍渲染图
4. 大象山舍社区景观示意图

1. New York Soori High Line
2. DAXIANG Render
3. DAXIANG Render
4. DAXIANG Landscape Design Sketch

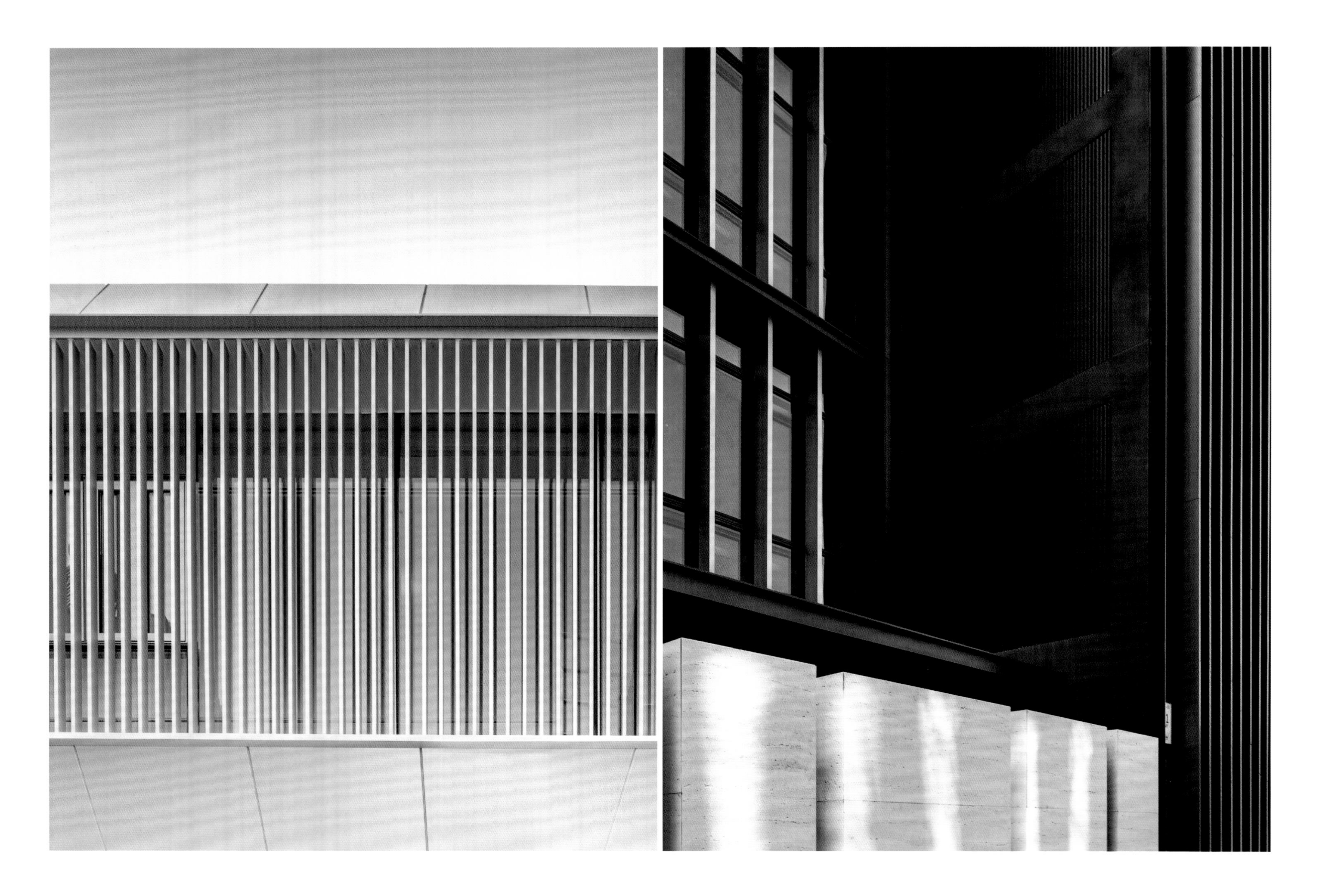

1 2 3

1—3. 大象山舍实景图

1—3. DAXIANG at Different Viewpoints

PART C Green Building
绿色建筑

07 Shenzhen Vanke HQ

深圳万科中心
水平摩天楼

建筑设计：史蒂芬·霍尔建筑事务所
项目地点：广东省深圳市盐田区大梅沙环梅路 33 号
设计时间：2009 年
完成时间：2006—2009 年
建筑总面积：120,440 ㎡
功能：办公楼、综合体、公寓、会议中心、酒店

Archi. design: Steven Holl Architects
Location: No. 33 Huan Mei Road, Dameisha Yantian District, Shenzhen, Guangdong
Design process: 2009
Completion time: 2006—2009
Total area:120,440m²
Program: Office, Complex, Apartment, Conference, Hotel

与其所在垂直的山地结构不同，深圳的城市结构正在向水平方向蔓延。深圳有很多的公园和绿化带。从市区向东走，穿过一条隧道，就可以看到坐落在深圳大梅沙海滨公园的“水平摩天楼”正俯瞰着眼前浩渺的大海。与以往建造出一系列单独的建筑形体，包括酒店、公寓、办公楼等的传统方案不同，霍尔事务所新的方案构想是一个漂浮建筑。这是一个属于未来的构想，它的水平几何形态连贯在一起，犹如海平面将各种功能空间提升到更开阔的视野。万科的标志性形象可以随景观绿地的最大化而最大化。

作为深圳城市片段，此设计提出一个新的典范：水平摩天楼。这个概念包括飘浮的水平杆状空间，用以化解建筑形式和功能使用之间的直接关系，这将给地面更多的活力。在这个项目中不需要设计各种不同功能来满足这个城市片段的复杂需求，地面层多元的日常生活可以在功能单元中不断改变和演化。这些单元和周遭活动之间的多孔穿透性是非常重要的。由于主楼飘浮在空中，这些地面出租的空间可以让租户使用当地的自然材料自己建造，例如竹子、茅草屋顶等，并且可以提供紧密多样的使用性，使其具备很大的可变性和灵活性。

盘旋在独创的“海水涂鸦”花园上空，在办公室、公寓和酒店等建筑物之间温和地碰撞，就好像它们一度曾漂浮在较高的海面上，如今那个海面已经退去，留下它们屹立在犹如玻璃或珊瑚般的基座上。这种项目是在中国，乃至世界任何其他地方都从来没有出现过的。作为一个热带的、可持续的 21 世纪构想，它融合了几项新的可持续发展方向：飘浮的建筑体创造了自由、灵活有遮盖的景观绿地，并且让海风和陆风穿透基地。利用中水系统运作的矩形水景池，将冷能向上辐射到彩色的铝制建筑底面再反射下去。可动式外遮阳表面使用特殊复合材料，保护内层玻璃减少太阳能负荷及风力冲击。可转动式悬挂立面外遮阳系统不会阻挡窗外的海景及山景。利用太阳能的除湿和冷却系统经由特殊的“屋顶阳伞”形成了有遮阳的屋顶景观。这个防海啸的盘旋式建筑创造了一个多孔的微型气候和庇荫自由景观绿地。

这将是中国南方第一个得到 LEED 白金认证的建筑。在霍尔的规划设计中，通过在景观上增加建筑，将原有的项目占地面积 6 万平方米地块拓宽成 7.5 万平方米的绿色空间。由此实现了绿色空间的最大化，并将其开放成为城市主要的公共空间之一。目前，已经有来自附近社区的居民在这里聚集活动。

在悬浮的建筑主体之下，霍尔事务所创造了具有 360 度视角的“深圳之窗”。大多数建筑物都有 4 个立面和一个屋顶——该项目的正面即是它第六个立面，而这一立面是直接朝向地面的。对于一座水平摩天楼来说，这个立面才是最重要的。在“深圳之窗”，你可以从这里沿着走廊向下漫步,然后观看整块场地的 360 度全景。该建筑有 20 米高，而当你站在其中时，它的体量似乎就要消失了。建筑物的阴影总是在随着日照不断变化，再加上徐徐吹来的海风，建筑下方的景观区域就形成了一个小范围的气候带。与此同时，第六个立面的倒影投射在建筑下方的水池中，发出熠熠的光彩。

1. 霍尔手绘草图

1. Painting by Holl

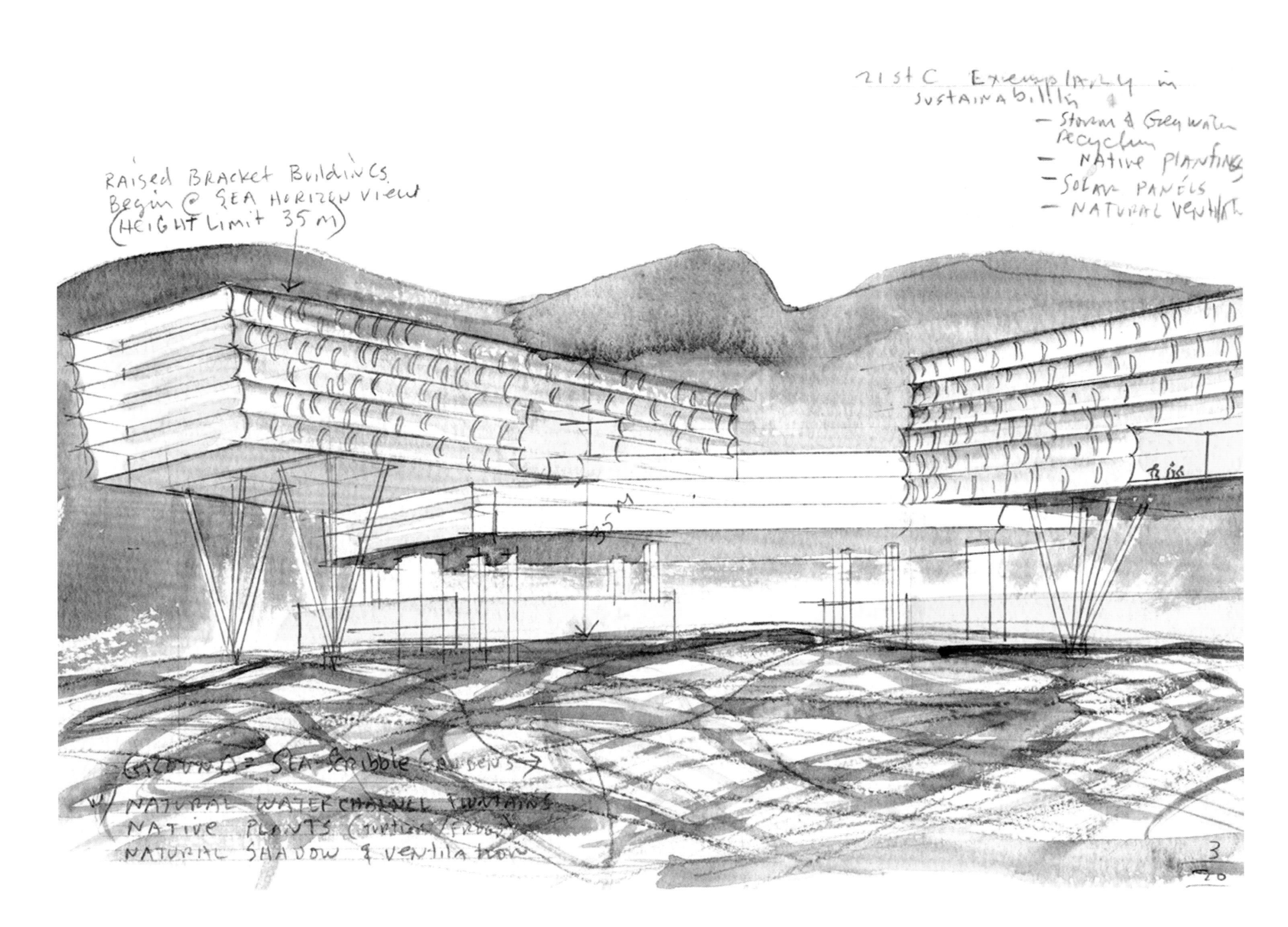

Shenzhen's urban fabric is spreading horizontally against a mountain backdrop. There are a lot of parks, planning with greenery. Moving east after you pass through a tunnel on a bay you arrive at the site of the "Horizontal Skyscraper" overlooking the sea at Dameisha.Steven Holl Architects is committed to visualizing new floating building, an idea of future, connected in a horizontal pattern of geometry, like a rising sea level to elevate various functional spaces to the open view, instead of building a series of single buildings including hotel, apartment, and office buildings. The symbolic image of Vanke can be maximized along with the large landscape green space.

As a part of Shenzhen, the design presents a new role model; a floating horizontal truss structure space, defusing a direct relationship between building forms and functional utilities as a way to add more vigor and vitality to the ground floor. This particular project does not need to design different functions to meet the complex requirements of this area, while the fact is that the diversified daily life in the ground floor can be changed and evolved gradually in these functional units. The perforated penetrability between these functional units and its surrounding activities is much important. Due to the main building is floated in the air, other spaces on the ground floor for rental can be used by leasee, who can use local national materials, for example, bamboo, thatch roof, to create their own space, together with the diversified functions of use, making the space with the possibility of variable and flexible.

Hovering over the garden with originality style of "sea scrawl", the collision between buildings including offices, apartments and hotel, likes they are at one time floating on the higher sea level. Nowadays, the sea water has receded, the left buildings stand on the glass and coral style basement. This project is the first of this kind in China (or anywhere). As a tropical strategy and a sustainable idea in 21th century, it integrates new several sustainable development aspects: the floating building creates a covered landscape green space with free and flexible style, making the seal breeze and land breeze across, even penetrate the base. Rectangle waterscape pool equipped with reclaimed water system can make the cold energy radiate upward to the color aluminum building bottom, and then reflect it back. Special composite materials are applied to movable external sun-shading surface to protect the inner glasses from the solar load and wind impact. Rotated suspended elevation external sun-shading system can't obstruct the seascape and mountain views outside of window. The dehumidification and cooling system with solar panels via special "roof parasol" forms a sun-shading roof landscape. The building is a Tsunami proof hovering architecture that creates a porous micro climate and shading landscape green space.

This will be the first certified LEED platinum building in Southern China. The site is 60 000 square meters. Holl' s design, on 60,000 square meters, has the building above the landscape, and as a result the site have 75,000 square meters of green space. The design maximized the green space, and open this as a major urban public space. Already people in the surrounding community push baby buggies there, walk their dogs and play.

Holl's design had invented a "Shenzhen Window" a 360° window, which hangs below the soffit. Most buildings have four elevations and a roof—this project's face is in the sixth elevation, the sixth elevation being the underside. For a Horizontal Skyscraper, that's the primary elevation, the sixth elevation. In the Shenzhen windows, you can just be walking along the corridor and drop down and get a 360° view around the site from here. The building, 20 meters up, seems to disappear when you arrive at the site. With sun shining,the shadow is always changing. The sea breezes come under, and you have a real microclimate landscape below. The water gardens reflect the colored "6th elevation" of the building floating above.

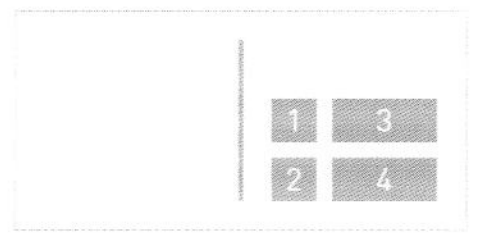

1. 视觉最大化示意图
2. 尺度对比分析
3. 流线动向轴测图
4. 景观最大化示意图

1. Maximize Views
2. Size Comparison
3. Path Diagram
4. Maximize Landscape

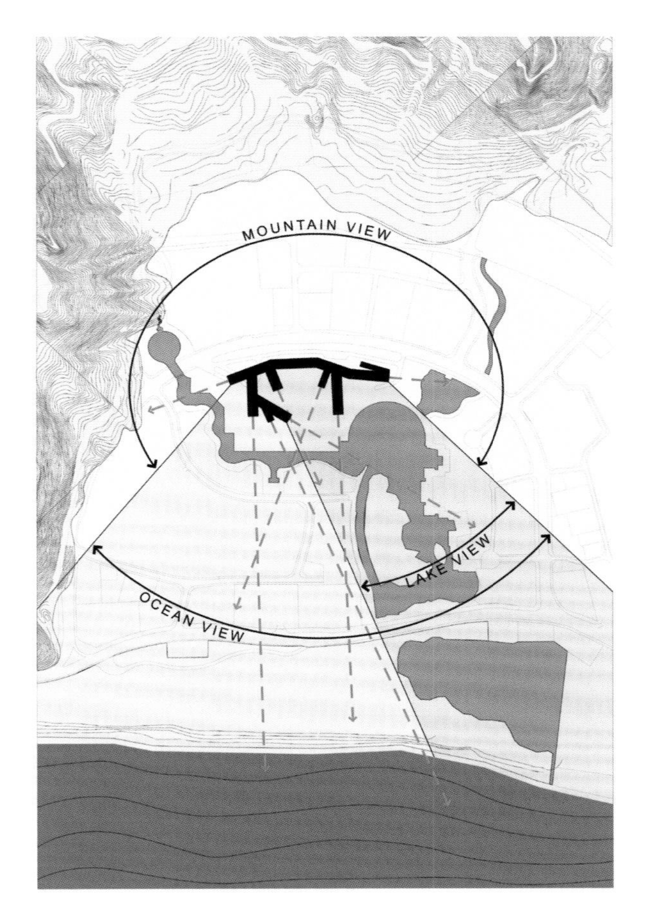

MAXIMIZE VIEWS

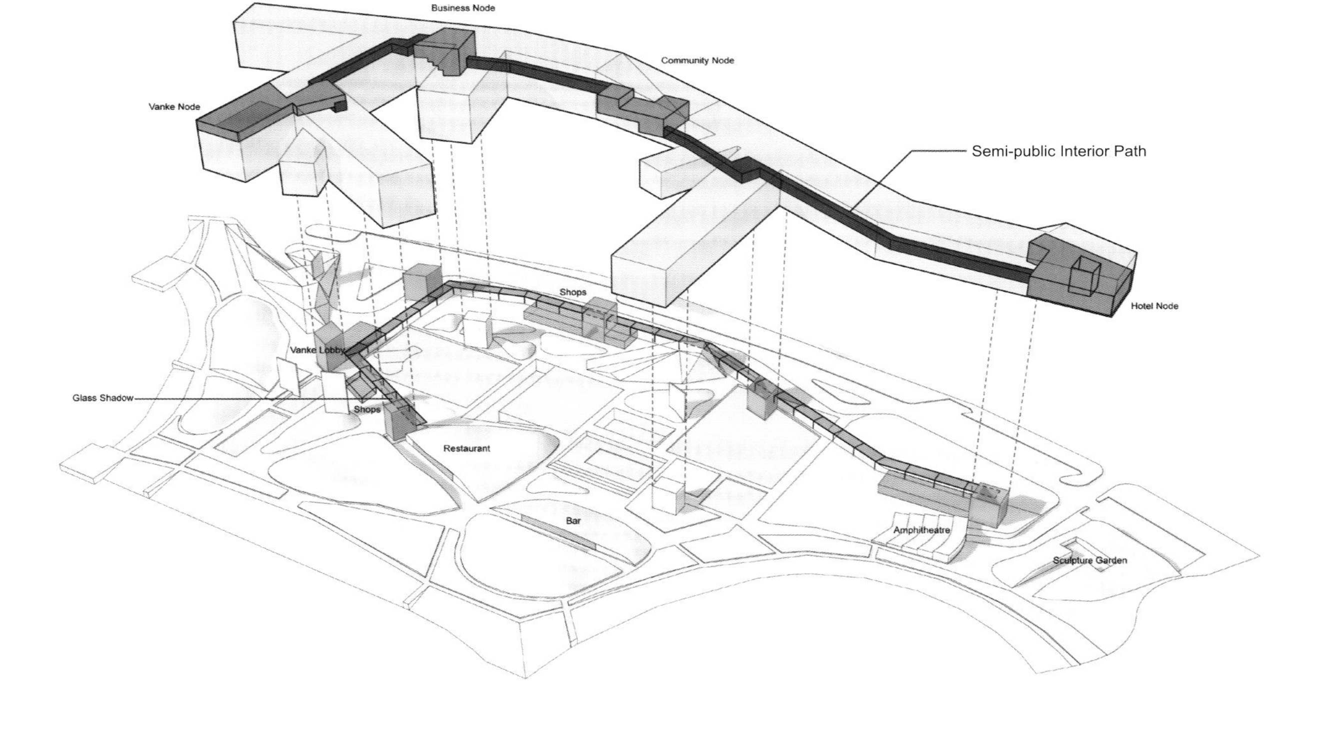

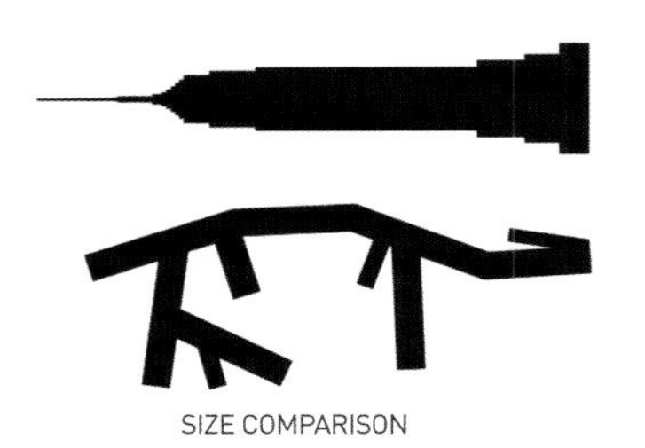

SIZE COMPARISON

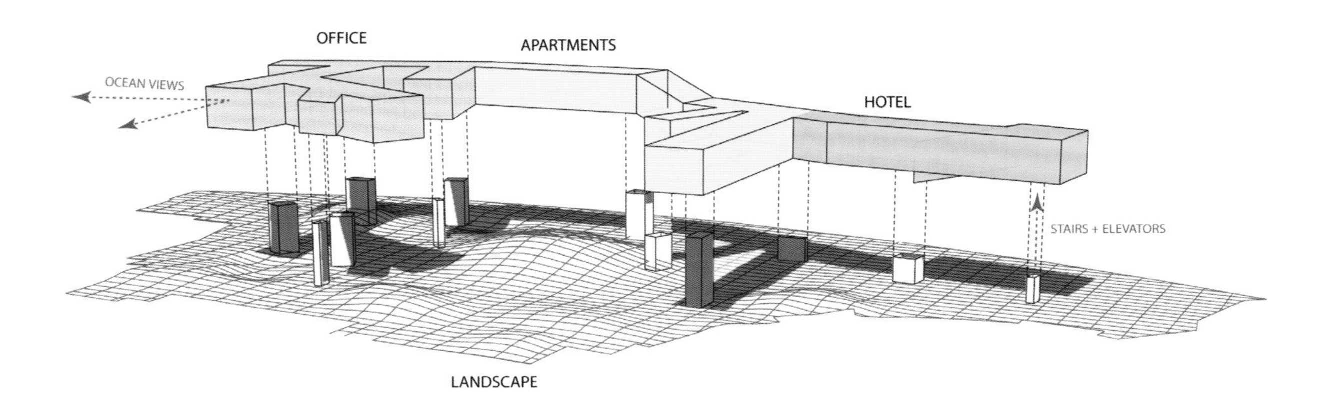

“鉴于多层面的不确定性，在 21 世纪，我们必须学会在充满歧义和疑问的环境下工作，在各类实验性的建筑设计中，相比于建筑本身，建筑内空间的形成过程才是我们的首要目的。”

关于建筑师

霍尔的建筑风格激扬而不张扬，既有天马行空的诗意，又含深刻切实的哲理，这无疑与他早年的不顺和沉寂有关，也是他不懈追求艺术和卓越的反映。他学徒时代（1966—1976）在西雅图、罗马、旧金山和伦敦度过，创业之初（1976—1993）则在纽约教书糊口，虽默默无闻，但受梅洛庞蒂启发而从类型学转向现象学，逐渐自成一格。他相信观念驱动设计，所以他的作品都其来有自，或源于神话隐喻、科学概念，或源于音乐舞蹈、绘画文学。比如他在国际比赛上的成名作，赫尔辛基的奇亚斯玛当代艺术博物馆，其名称和设计就源自视神经的交叉。不过，霍尔同意维特根斯坦没有现象学而只有现象学问题之说，认为观念流动无常，理应不断追问超越。所以他也强调地方特性和场所体验：如果说他的北京当代 MOMA 是把城市变得立体，让社区内部互动，那深圳万科中心则是让城市回归平面，把空间留给公众。这些都体现了霍尔对新型建筑和开放都市的探索，也显示出他对中国社会变迁和可持续性发展的思考。如今霍尔工作室在北京和纽约的团队连轴运转，而他在忙碌中却愈发从容，随时画画或阅读，在安静中酝酿新作的诞生。

ABOUT THE ARCHITECT

Holl has always been a poetic designer that carried out projects that are exciting with deep practical philosophical concept without overtly aggressive display. It certainly relates back to his early years, which was quiet and rather unsmooth. Holl's constant pursuit in perfection and art had started since his start of the career. From 1966 to 1976 Holl spent his apprentice years in Seattle, Rome, San Francisco and London. His early career began with teaching in New York, completely unrecognized by the architecture realm, Holl was quietly inspired by Merleau-Ponty and gradually established his style from typological perspective to phenomenology. Strongly believed in concept-driven design, Holl's architecture works can all be back-traced to inspirations from mythological metaphors, science concepts, music, dance, paintings or literature. For instance, his famous masterpiece, Museum of Contemporary Art KIASMA in Helsinki, involves the idea of intertwining of optic nerves and reflects in both the name of the building and the design of architecture masses. On the other hand, Holl follows the philosophical idea of Wittgenstein's, he believes that perspectives and concepts can change unpredictably and designs should strive for constant questioning and exceeding. As a result Holl specifically encourages site features and experiences. The project "Linked Hybrid" MOMA in Beijing resulted a 3-dimensional city with excellent community interaction; while the "Horizontal Skyscraper" Vanke Center in Shenzhen had brought the city-scape down to a 2-dimensional scale for a better public-space experience. Both projects reinforce Holl's exploration in new architecture types and the idea of open-city, and continuously reflect his thinking in social reform in China and sustainable development. Holl's offices, both in Beijing and New York, are bustling with works, while Holl retains an easy life of painting and reading, subtly brewing new ideas in peace.

史蒂芬 · 霍尔
Steven Holl

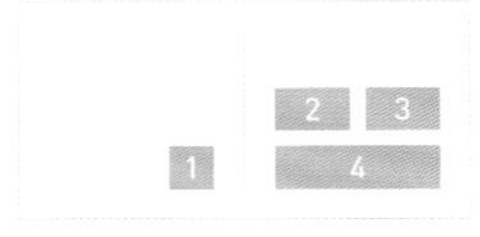

1. 普林斯顿大学路易斯艺术中心
2. 模型示意
3. 视觉语言：水平几何链接的架空建筑
4. 模型立面

1. Lewis Arts Complex, Princeton University
2. Model
3. Visual Language: Horizontal Elevated Geometric Connection
4. Model Elevation

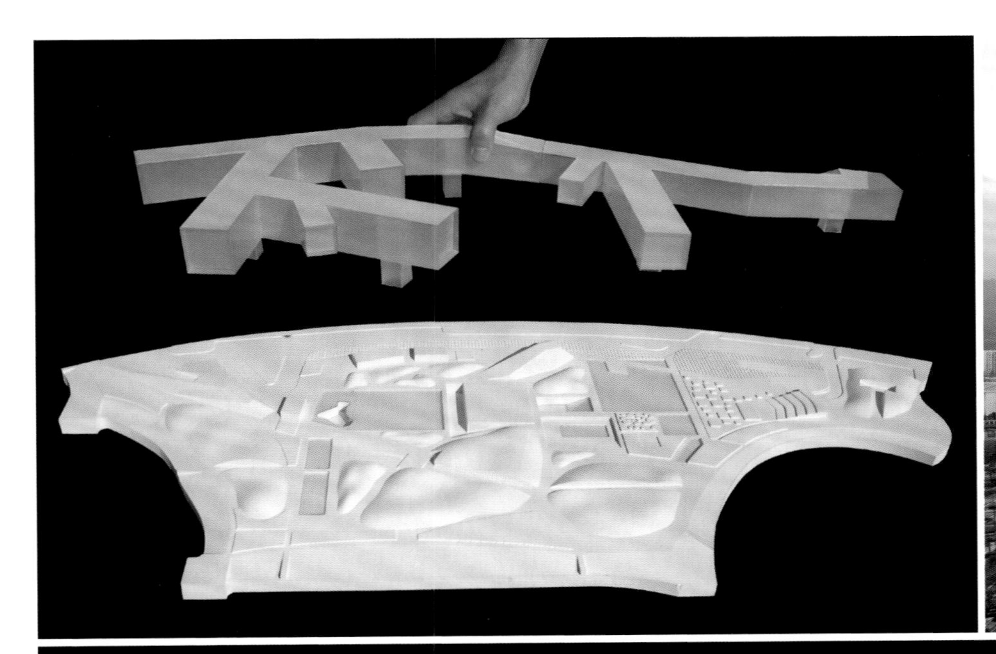

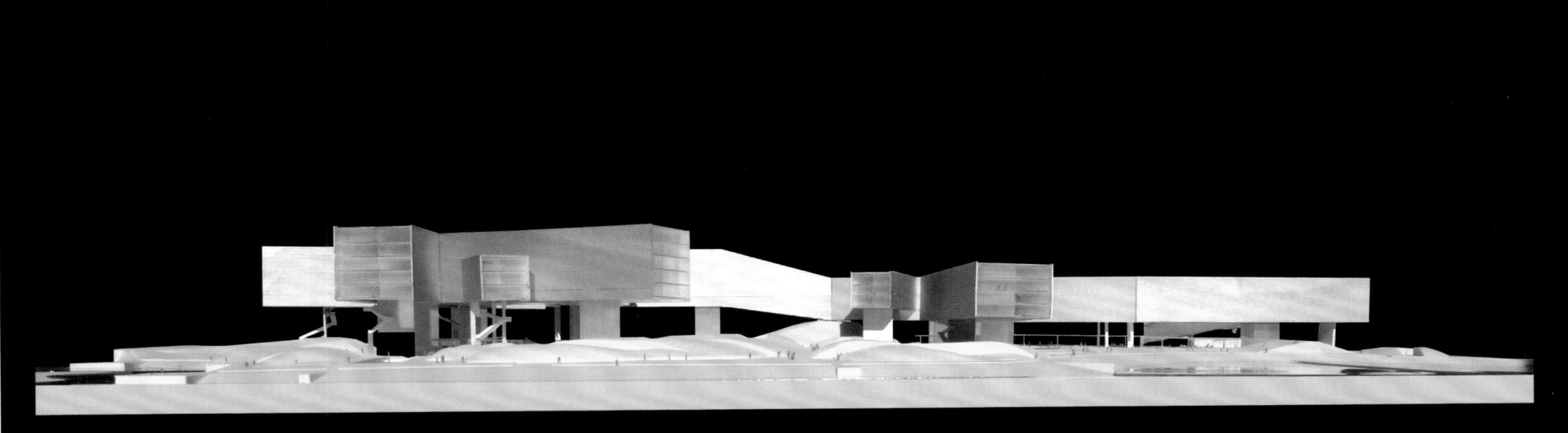

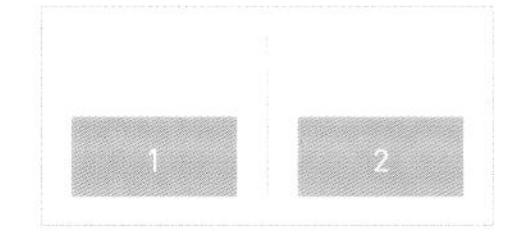

1. 悬浮结构的下方景观
2. 建筑下方的主立面形成了"第六个立面"

1. The Underside of the Floating Structure
2. The Underside Became the Main Elevation: "Sixth Elevation"

HILIPS
Vanke

08
2010 Shanghai Expo Vanke Pavilion

2010 上海世博会万科馆

在高高的麦垛，看冬之雪，秋之收获

建筑设计：多相工作室

项目地点：上海市黄浦区半淞园路 2049

完成时间：2010 年

建筑总面积：3,309 ㎡

功能：展览

Arch. design: Duoxiang Studio

Location:No.2049 Ban Song Yuan Road, Huangpu District, Shanghai

Completion time: 2010

Total area: 3,309m^2

Program: Exhibition

2010 上海世博会万科馆昵称为“麦垛”。在设计竞标阶段，基于建筑及建材生命周期 CO_2 排放量的研究成果，“多相”工作室选用秸秆板作为结构材料，并以材料→结构→空间的推演为基础生成建筑的形态。最终实施的“麦垛”由表皮为秸秆板的三个正圆台与四个倒圆台交错组成，它们围合而成的半室外空间四周通透，顶部通过透明采光膜连成一体。圆台内部是独立的展厅与后勤办公空间，建筑外部环绕着景观水池。

整个建筑的空间分为两种形态：圆台内部空间，是封闭的、静态的、单纯的，功能是展厅、演播厅和后勤服务；圆台外部空间，是开放的、流动的、复杂的，作为中庭，功能是观众等候参观、问询和购买纪念品。空间内部倾斜的墙体不提供垂直参照，这对人的平衡感产生挑战，会使人上下左右地看这个建筑，以寻找并确认建筑的稳定。正、倒圆台的倾斜墙体彼此部分遮挡、部分展现，加之曲面的连续，使得中庭成为一个随着人的行经持续变化着的空间，让人产生探寻的渴望。

世博会期间真正能进入万科馆内参观的人是少数，因此多相希望形成一种可以和更多人分享万科馆的开放性的建筑。于是多相没有封闭中庭空间，并且用阻隔路径但不阻挡视线的水面围合建筑，而不是围栏、围墙，这样就可以使路上的行人不用进入万科馆就可以看到建筑内部，甚至看穿建筑。

多相工作室将占建筑总面积 1/6 的中庭设计为开敞的空间，完全不使用空调。中庭在 4 个方向均有开口，由于中庭周边的倒圆台形成的空隙都为上小下大，建筑外部的气流在经过建筑的时候，上部的气流被建筑的形体压向下部，从而使靠近地面的风速提高，人会感到更凉爽。中庭上部的 ETFE 膜气枕与女儿墙顶留有空隙，ETFE 膜在阳光照射下温度升高，可以加热中庭顶部空气，使顶部空气温度高于地面空气，实现热压通风。

通过自然材料（秸秆板）、自然采光、自然通风降温这几个方面的设计表达对自然的尊重，也期冀万科馆的建筑设计可以唤起人们接受、欣赏、尊重自然的观念与信心。

新秸秆板的自然纹理和金黄色泽都会让人感受到生命的健康与丰盛，但如同任何生命都会衰老死亡一样，秸秆板的色泽也会随着时间的推移而灰变。多相希望通过这种自然的蜕变可以传达一个观念，即如果人们尊重自然的应有状态，就会减少与自然的无谓对抗。

秸秆板的质感吸引了很多观众去触摸，因为这种不常见的材料超出了视觉固有的经验，他们需要用触觉提供更多的信息，以确认这种材料。这种激发视觉之外的感官体验的设计（包括对人的平衡感的挑战）对多相来说非常重要，如果缺少就会将建筑变为单纯的视觉经验，那么体验建筑和看建筑照片之间的差别也就不大了。

1. 总平面图
2. 剖面图
3. 立面图
4. 平面图

1. Masterplan
2. Sections
3. Elevations
4. Plan

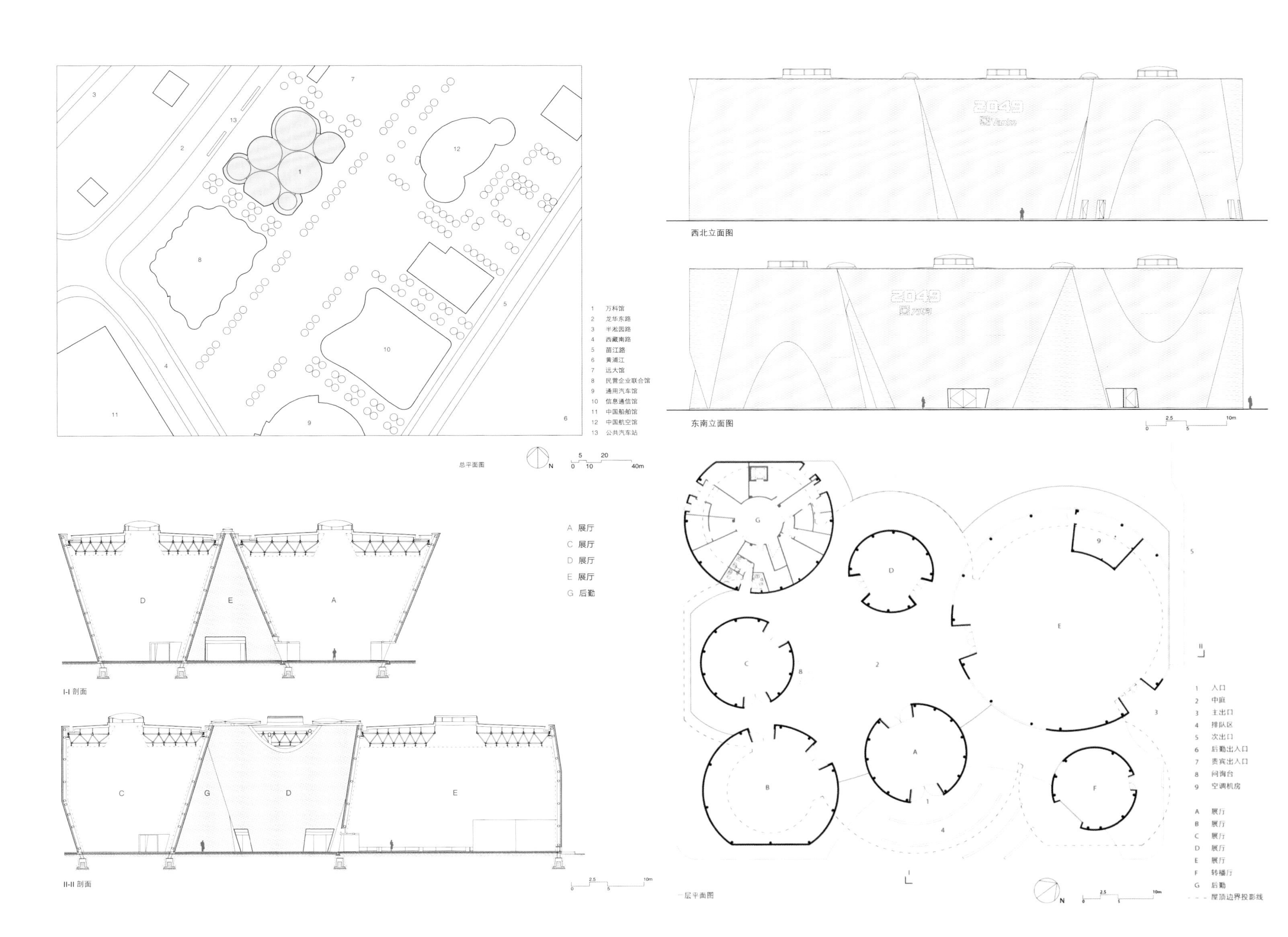

The Vanke pavilion featured at the 2010 Shanghai World Expo was given the nickname "straw stack". During the design competition stage, the conception of the pavilion was based on research findings derived from the life cycle of CO_2 emissions from building and building materials. Straw board was selected as the structural material, and the architectural form was generated based on the deduction of the material, structure and space in that order. The final implementation of the "straw stack" is based on a surface structure that is composed of three concave-up cone-shaped platforms and four concave-downwards-facing platforms that are interlacing. The result is that the semi outdoor spaces surrounded by them are permeable, and the top is connected by a transparent lighting film. The inside of the circular platform is an independent exhibition hall and logistics office space, surrounded by landscape pools.

The space of whole construction is divided into two kinds of form: the space inside cones is enclosed, static and simple, with functions of exhibition, media studio and office; the space outside cones is open, fluid and intricate, as a atrium with functions of waiting, inquiry and souvenirs purchase for visitors. Tilted walls without vertical reference challenges the balance sense of man, so that people will look the construction up and down to find and confirm the stability of buildings.The tilted walls of upright and reversed cones partially sheltering and revealing from each other, together with the continuous arc surface make the atrium a space changing continuously with people passing, attractive to explore.

Since only minority of people could enter Vanke Pavilion during the Expo, Duoxiang hopes to form an open construction that can be shared with more people. We haven't shut the atrium but used pool not fence or wall to enclose the buildings that blocks path without obstructing sight, then the pedestrian could see the inside building even see through buildings without entering Vanke Pavilion.

Duoxiang studios designed the atrium that occupies 1/6 of the total building area as an open space without using any air-conditioning at all. The atrium has openings in all four directions. Because the space formed by the inverted terrace around the atrium is smaller above and larger below, the air flow from the outside of the building is guided downwards by the shape of the building during the process of passing through the building, so that the wind near the ground level is enhanced and people feel cooler. There is a gap between the ETFE film air pillow and the top of the parapet in the upper atrium. The temperature of ETFE film rises when under the sunlight, which can heat the air in the top of the atrium, and make the air temperature around the top of the atrium higher than that of the ground level. The result is the realization of warm pressure ventilation.

Through the application of natural materials (straw board), natural lighting, and natural ventilation and cooling, these myriad aspects of the design are an expression of respect for nature, yet they are also an expression of hope that the architectural design of the Vanke pavillion can inspire people to accept, appreciate, and respect the concept of nature and to place greater trust in it.

The natural texture and golden color of new straw board have people feel the health and abundance of life, however the color of straw board will change grey as time goes on like any creature getting old to death. Duoxiang hopes to convey a concept with the natural fading, if people respect the due state of nature, the unnecessary confrontation with nature will be lessoned.

The texture of straw board appeals people to touch, because the unusual material is beyond the inherent visual experience, they need more information by touch to confirm it. The design to stimulate the sensory experience beyond vision is very important, without it the construction will be a pure visual experience, if that to experience the building will not be much different from looking at pictures.

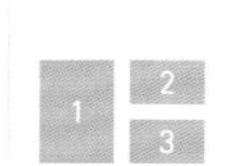

1. 在“麦垛”间穿行
2. 在“麦垛”内仰望
3. 感受“麦垛”外墙材质的人群

1. Walking through the Straw Stacks
2. Looking Up the Skylight Inside the Straw Stack
3. People Feeling the Material of the Exterior Surface

“我们的理想在于从过程中发现每一种‘相’的价值并发展这些价值。”

关于建筑师

“多相”暗示的是多种角度和状态，然而多相工作室有的不仅仅是多元开放的自由，还有志同道合的默契。陈龙、陆翔和贾莲娜是清华同学，研究生毕业后在非常建筑工作时遇到胡宪。四人受张永和启发而关注建筑本体、现代主义和中国性，并建立了平等友好的工作关系，最终基于共同的历练和价值观而在 2006 年合伙成立多相，在“怀疑”和“构建”中独立发展，“真实地触碰社会”，与业主一起对话和成长。从 798“案艺术实验室”这第一个项目到与“素然”的长期合作，多相的多数工作来自私人委托，项目虽小，但四人发挥集体智慧，乐在其中，和而不同。他们首次被邀标做的开发商项目就是上海世博会万科馆，一向重视建筑材料的多相因其使用秸秆板的低碳方案胜出，从而完成了他们里程碑式的转变之作，从简单的概念思维转向场所和生活体验，转向后来“石岛山居”时的通透成熟。可贵的是，多相“不会因为经济舍弃自己对设计的追求”，也从不“滥用风格”，用贾莲娜的话说，他们是通过“认真地生活产生风格”。

ABOUT THE ARCHITECT

"Duoxiang" implies a multiple status and perspectives, the design studio presents a diverse and open working environment with remarkably cohesive team. Chen Long, Lu Xiang and Jia Lianna were colleges from Tsinghua University, they worked at Atelier FCJZ and met Hu Xian after master studies. The four had inspired by Zhang Yonghe's idea of form, modernism and Chinese-featured design and established an equal-friendly working relationship. They had agreed on a mutual experience and value system and founded Duoxiang in 2006, with a mission statement of "independent development in questioning and constructing with true understanding of the real society". Maintaining a conversational-based development with the owners, Duoxiang had accomplished a series of private projects from "& Art Lab" in 798 to long-term cooperation with ZUCZUG. The team-of-four had enjoyed a congregated intelligent working method with fun and diversity. Shanghai Expo Vanke Pavilion was Duoxiang's first inviting-bid project by the developer, their low-CO_2-emission strategy with straw board material had emerged as the best candidate, the project hence became their mile-stone accomplishment. Duoxiang had tranformed from simple conceptual thinking to site and life experience design, and reached its climax in Shidao Resort. Duoxiang would not abandon their design ideas due to economic constrains, their consistent pursuit in "cherishing life and deriving style" (quoted from Jia Lianna) and scrupulous in design manner are commendable.

多相办公室
Office of Duo Xiang

1. 石岛山居
2. 鸟瞰
3. 外景
4. 内景

1. Shidao Resort
2. Birdeye View
3. Exterior View
4. Interior View

1 2 3

1—2. “麦垛”内部
3. 室外全景

1—2. Interior of the "Straw Stack"
3. The Pavilion View from Outside

09

Field House

田屋余舍

明日的村庄

项目地貌典型独特，中部山地，四周平原，湖景环绕，有坐山望湖之景。旧有矿坑深入山体，湖波水域景观质量高，群落具有自然演替性和次生性。项目具备重要的鸟类资源，是鸟类东部迁徙通道的最重要地段，其亚热带季风气候使得现状地块内植被整体长势较好，并拥有极具观赏价值的黑松林。因此生态修复、合理建设以及保护性开发是项目整体规划的重要指导思想。

大岛北部拥有大面积的湿地，最为生态，与小岛联系紧密，可作为生态论坛主空间与主会场，有举办大型活动的承载力。大岛南生态景观资源种类丰富，西侧滨水尖地与大矿坑各有不同的风情，可提供不同的生态度假体验。小岛地势平坦，相对独立，生态敏感度比大岛更低，对人群活动有更好的承载力，更适宜对外公共活动的组织。

The project has a unique topography, the middle of the islands are hilly with excellent view of the lake and the peripheries are surrounded by plains. The hill is penetrated deeply by old mining pit. Local coenosis is featured with natural succession bio-habitat and secondary vegetation. Gongshan Island is famous for its bird resources as a key location on eastern migration corridor, the subtropical monsoon climate had made the current site with excellent vegetation coverage and Wayward Pines with great scenic value. Consequently the overall planning for the project is emphasized on ecological restoration, appropriate construction and protective excavation.

The northern side of the Big Island contains large area of wetland of the most ecological value, the linkage to the Small Island is close enough and the size is large enough to establish functions such as ecological forums and major conference hall. On the south side there are rich ecolandscape resources, the west of the site is a waterfront and the old mine pit, which is suitable in tourism of unique experiences.The Small Island, on the other hand, is relatively plain in topography and lower in ecological sensitivity, which is appropriate for exterior public activities for the travelers.

建筑设计：大舍建筑设计事务所、刘宇扬建筑事务所、直造建筑事务所

设计时间：2017 年

功能：建筑概念

Arch. design: Atelier Deshaus, Atelier Liu Yuyang Architects, Naturalbuild

Design process: 2017

Program: Architectural Concept

茶田余屋

围绕于茶田之中，具有传统江南农耕文化、田园文化的村落式精品酒店。

Field House

Boutique hotel in the tea field, village-style layout featuring traditional southern China garden and farming culture

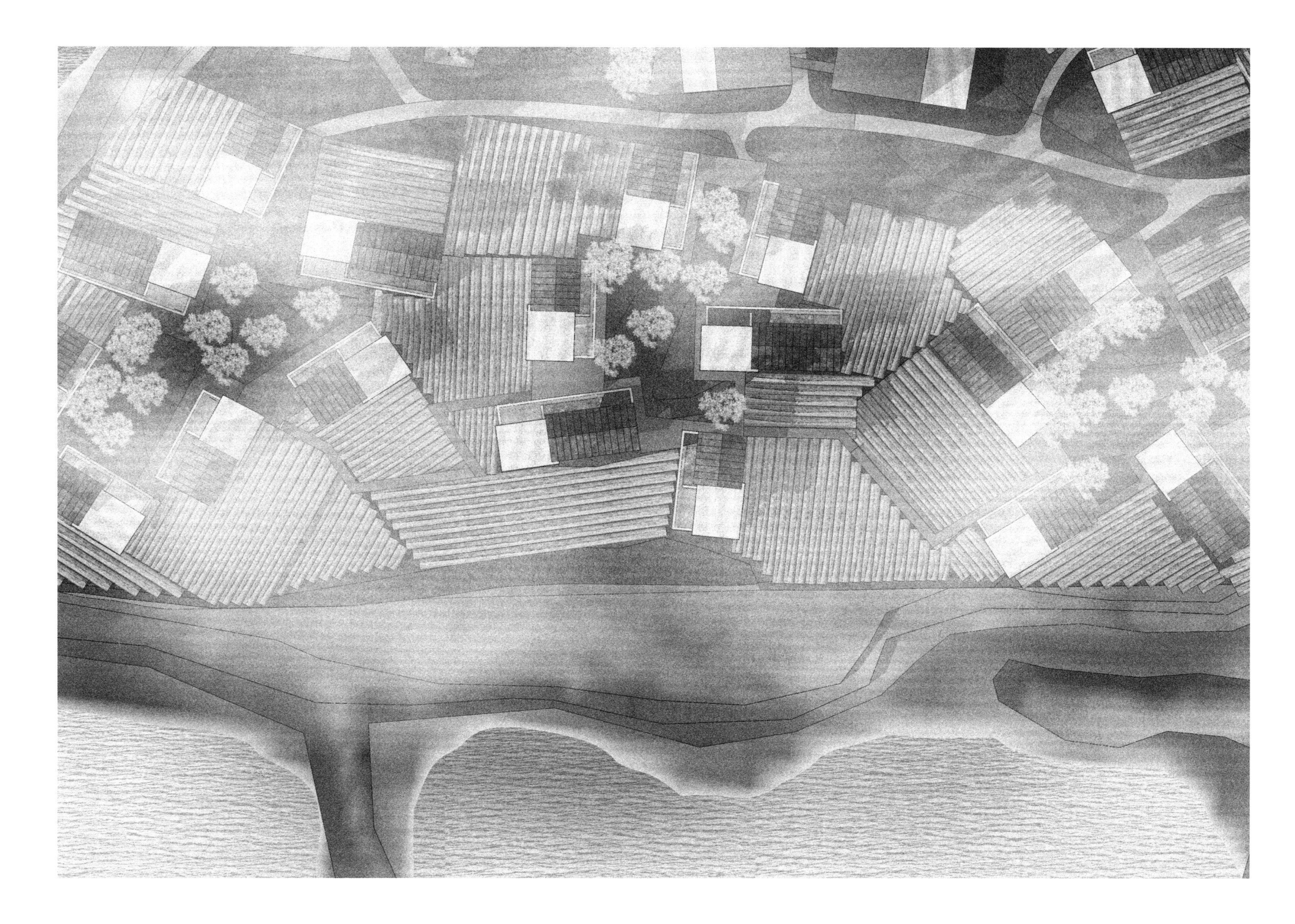

矿坑酒店

项目选址于面向湖景的矿坑遗址，为最大程度地实现建筑和湖景平山远水的互动，呼应苏州传统园林中对真山水的追求，设计在集中式与自然聚落式酒店布局之间取得一种平衡，既让住客感受到多层次的湖景，又通过建筑本身修复矿坑地景，使建筑成为岛屿景观的一部分。

Mine Pit

Mine Pit is located at the old minepit of the site facing directly to Lake. The design aims to maximize the interaction between the building and the landscape of Lake and mountain while echoing the pursuit of traditional Suzhou Garden style. The hotel layout is seeking a balance between the centralized design and scattered village style design, rejuvenating the minepit landscape with the architecture itself, hence the building becomes part of the island sight.

山水跌合

业主希望对既有建筑进行一系列的改造以及新建。对于既有建筑的改造，我们选择以创造跌落山水的方式回应这片历史地域，与烟波浩渺的湖景相映成趣；而面对开阔的新场地，散落式的村庄布局能更好地分散体量，合院的表达也在回归传统的空间意向；而洞石、竹编等材料的介入也是基于太湖民俗的转译。我们意在让这片风景秀丽的地方，为追寻回归本真、回归田园的城市来客创造一个惬意的现代归宿。

Landscape House

The owner wishes to have a series of reconstruction and new buildings at existing buildings. The reconstruction scheme for the existing building is a reflection of the drop-Shan-shui style, integrating the view into the hazy lake sight. The scattered layout is appropriate in diversifying the volumes at the new open area, and the court-yard pays respect to the traditional space idea. The typical elements such as Taihu stone and bamboo are also implemented to create a garden-like modern resting space for the tourist from the city.

“建筑是心灵与土地的连接。”

“设计源于生活、终于愉悦；重现看不见的光、被忽略的美，这就是设计的本质。”

“当设计中的功利主义 (utilitarianism) 走向极致，也就带来了诗意。”

关于建筑师

大舍

大舍建筑设计事务所（Atelier Dashaus）2001 年成立于上海，现任合伙人与主持建筑师分别为柳亦春与陈屹峰。大舍的建筑作品获邀参加了法国蓬皮杜中心、威尼斯双年展等高规格展览，获得美国 AR 新锐建筑奖、中国建筑学会创作金奖等重要奖项，在超过 26 个国家的建筑与时尚杂志上广泛发表。

Atelier Dashaus was founded in Shanghai in 2001 and currently led by associate Liu Yichun and chief architect Chen Yifeng. The architecture projects of DASHAUS were invited to high profile exhibitions such as the Biennale di Venezia, exhibitions at Centre Pompidou in France and awarded with AR Awards for Emerging Architecture and Golden Award from Architectural Society of China. DASHAUS's projects are widely published in international architecture and fashion magazines in over 26 countries.

刘宇扬建筑事务所

刘宇扬建筑事务所（Atelier Liu Yuyang Architects）成立于香港和上海，由著名建筑师刘宇扬先生主持，近年来研究及实践的方向侧重于城市的微观生态性、建筑的在地文化性和节能环保性措施。一直以来，在工作中追求真诚的、有意义的、可持续的建筑与环境。公司理念是扎根本土、面向世界，以研究为设计之本，通过高品质的设计为客户及社会创造更高价值。

Atelier Liu Yuyang Architects was founded in Hongkong and Shanghai and led by renowned architect Liu Yuyang. In recent years Liu Yuyang and his design team emphasizes on urban-micro-ecology, cultural aspects in architecture localization and environmental strategies in energy-saving. Atelier Liu Yuyang Architects strive for sincere and meaningful architecture with a sustainable design approach, with its root in China, the design studio opens up to the world and works attentively for high-quality projects.

直造

直造建筑事务所（Naturalbuild）由水雁飞、马圆融、苏亦奇于 2011 年在纽约成立，现立足于上海。直造积极地拥抱现实中的矛盾性与复杂性，同时以设计激发不同基因之间的共鸣（功能、场地、材料、地形、气候等）。面对每个项目的具体性和多变的限制，直造通过理性推演与物理研究的工作方式，来探寻建造中新的可能性，从而在当代生活中延展更广义的自然。

Founded in 2011, Naturalbuild is a multi-disciplinary architecture studio based in Shanghai led by Yanfei Shui, Yuanrong Ma and Yichi Su. Naturalbuild embraces the complexities and contradictions of the reality in China, while also initiates the resonance of various genotypes emerging from program, function, landscape, topography, materiality and form etc.
In respond to the specificity and the ever-changing constraints of each individual project, Naturalbuild designs through rational deductions and tactile studies to seek for new possibilities in practice that would expand the meaning of naturality in contemporary living.

1 2 3

1. 大舍 — 夏雨幼儿园
2. 刘宇扬工作室 — 滨江爱特公园
3. 直造 — 莫干山大乐之野庾村民宿

1. Dashaus — Xiayu Kindergarten
2. Atelier Liu Yuyang — Riverfront Aite Park
3. Naturalbuild — Lost Villa Boutique Hotel in Yucun

柳亦春，陈屹峰
Liu Yichun, Chen Yifeng

刘宇扬
Liu Yuyang

水雁飞、马圆融、苏亦奇
Shui Yanfei, Ma Yuanrong and Su Yiqi

吴从宝
Wu Congbao

PART D Support To The Minds
精神配套

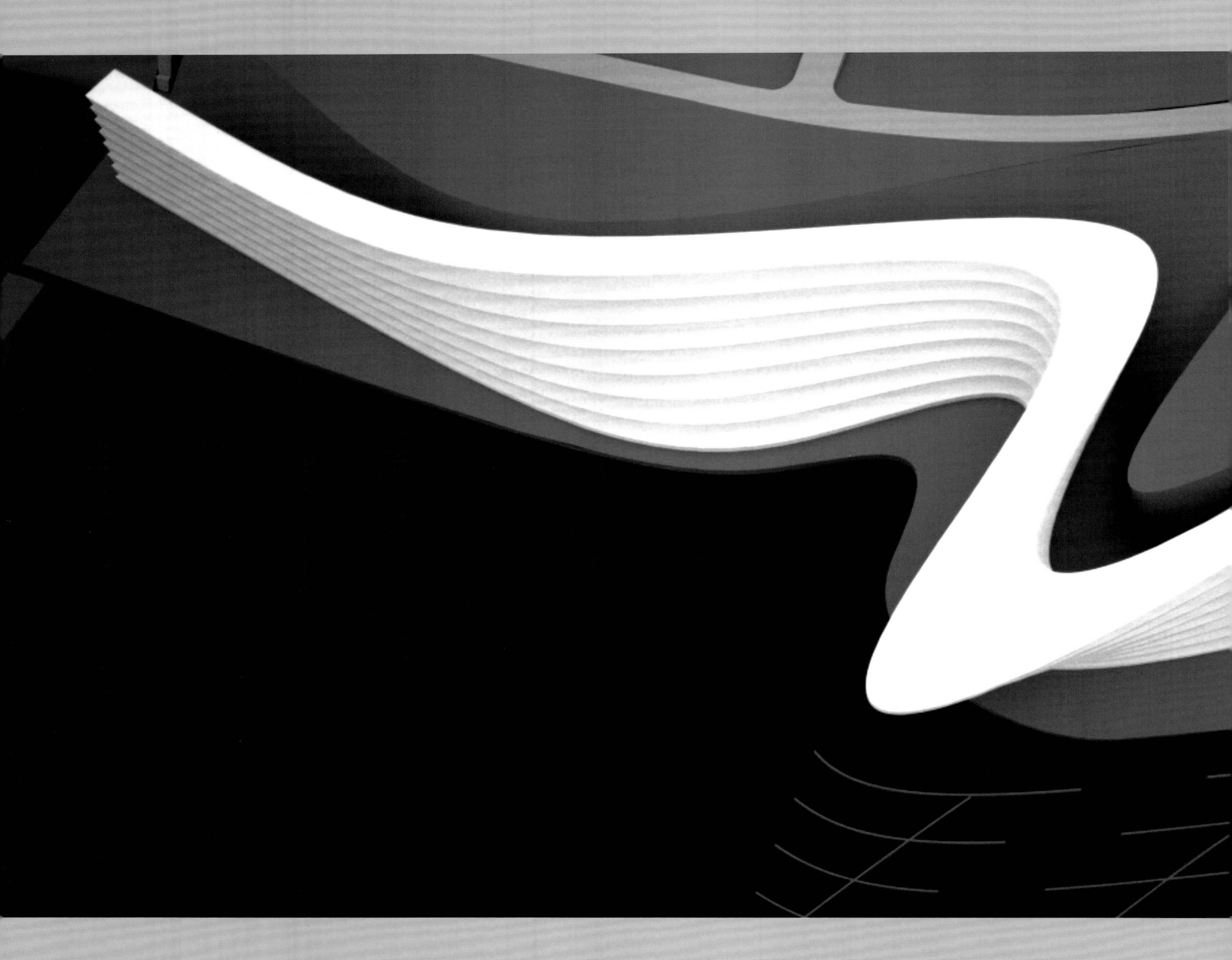

10 Shenzhen Long Cheer Yacht Club

浪骑游艇会

永远流动的曲线

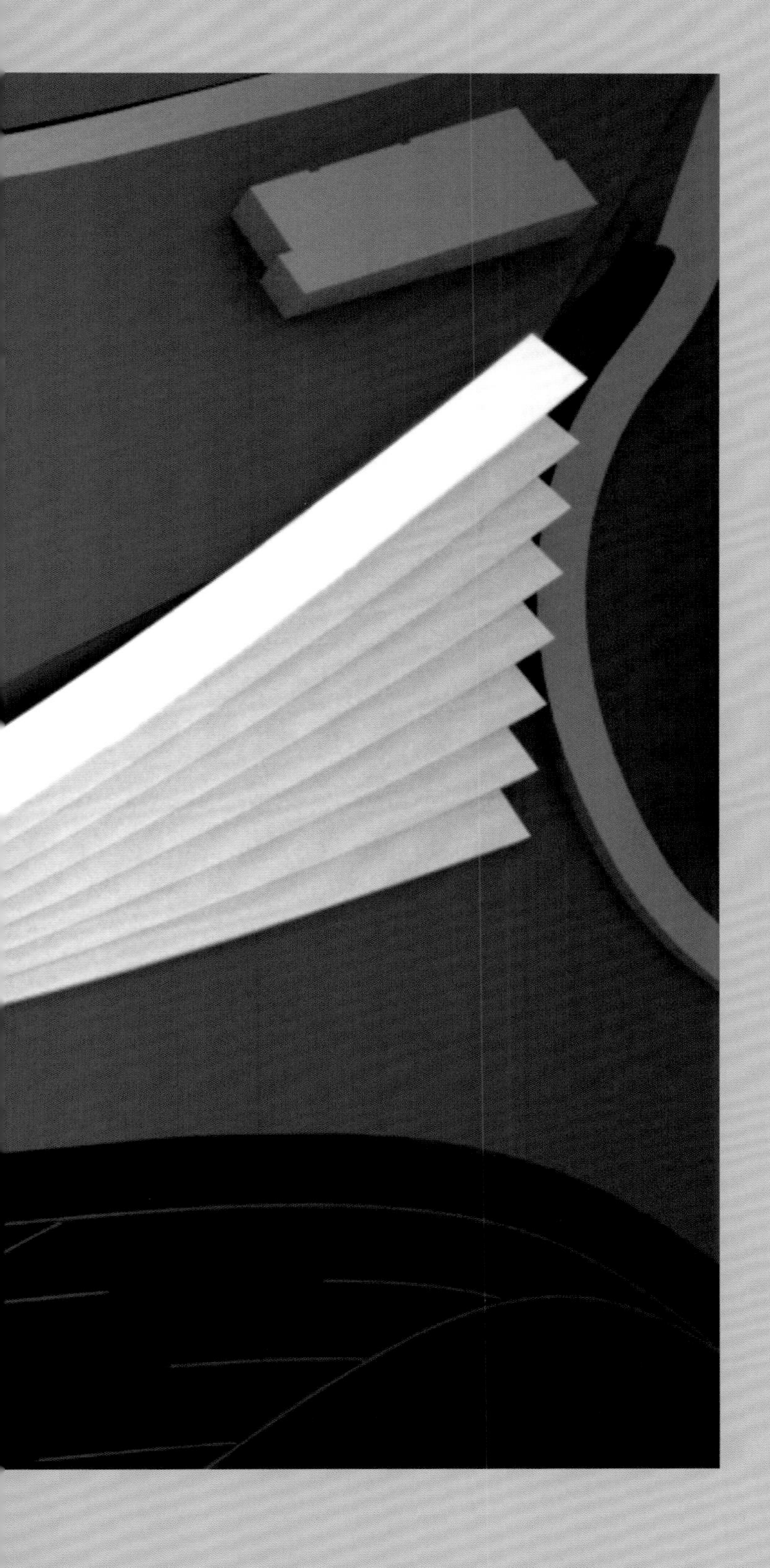

建筑设计：扎哈·哈迪德建筑事务所

项目地点：深圳南澳的大鹏半岛

设计时间：2010 年—待建

建筑总面积：30,000 ㎡

功能：俱乐部

Arch. design: Zaha Hadid Architects

Location: Nan'ao Peninsula, Dapeng District, Shenzhen

Design process: 2010—Present

Total area: 30,000m^2

Program: Club

扎哈 • 哈迪德建筑事务所创建于 1979 年，40 年来一直处于建筑、艺术和设计领域前沿。事务所因其独具活力和创新的建筑设计而享誉全球，这都基于扎哈 • 哈迪德对于都市生活、建筑和设计的改革性探索和研究。她因对于建筑领域的杰出贡献获得 2004 年普利兹克建筑奖。她的设计兴趣在于探索建筑、景观与自然地势之间的完美契合。其作品融合了自然环境与人工系统，试验性地运用尖端建筑技术把前卫的设计理念变成现实。这种建筑手法往往会创造出让人意想不到且充满活力的建筑形态。她的作品极具影响力，譬如美国辛辛那提“罗森塔当代艺术中心”以及中国广州大剧院，都被认为是体现新型空间概念和大胆视觉形式的设计，颠覆着人们对于未来建筑的构想。

Zaha Hadid Architects (ZHA) was founded in 1979 and has been at the forefront of architecture, art and design for 40 years. The practice is known internationally for dynamic and innovative projects which build on Zaha Hadid's revolutionary exploration and research in the interrelated fields of urbanism, architecture and design. In 2004 she received the Pritzker Prize in recognition of her outstanding contribution to architecture. Working with senior office partner Patrik Schumacher, Hadid's interest lies in the rigorous interface between architecture, landscape, and geology as her practice integrates natural topography and human-made systems, leading to experimentation with cutting-edge technologies. Such a process often results in unexpected and dynamic architectural forms. Previous seminal buildings such as the Rosenthal Center for Contemporary Art in Cincinnati and the Guangzhou Opera House in China have also been hailed as architecture that transforms our ideas of the future with new spatial concepts and bold, visionary forms.

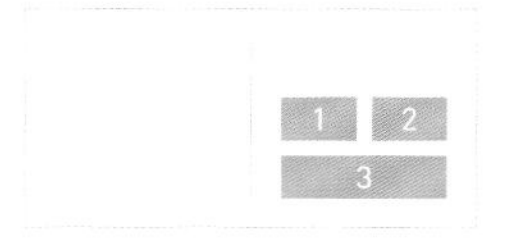

1. 总平面示意图
2. 曲线流动模型示意图
3. 数字化建模过程

1. Masterplan Diagram
2. Dynamic Curve Model
3. Parametric Form Generation Process

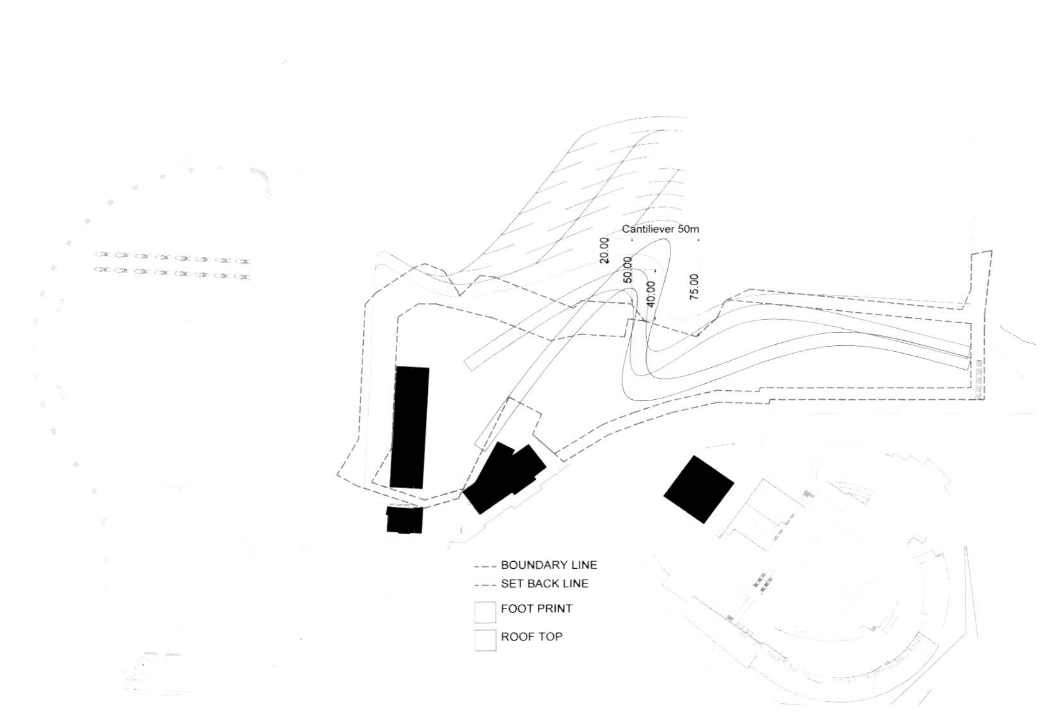

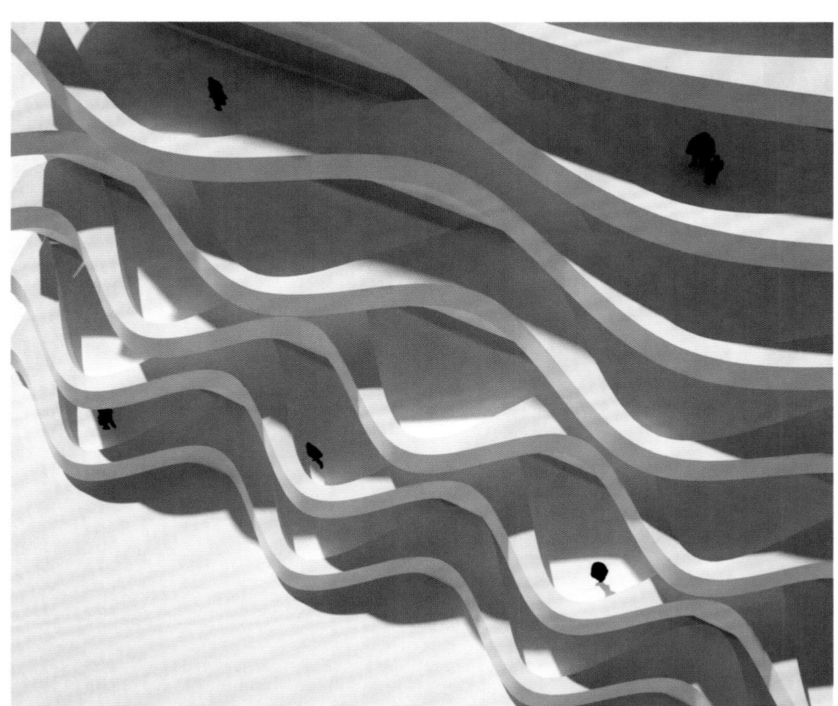

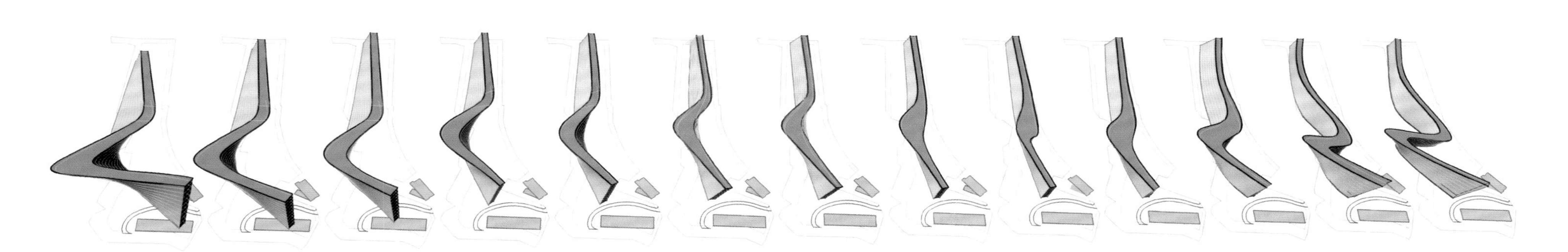

“我是一个女人，一个阿拉伯人，更是一个建筑师。从根本上，我是一个乐观主义者，并坚信自己的工作，最终也能排除万难。”

关于建筑师

“建筑是简单的”，一位仰慕扎哈的女性说，“看似结构复杂，但其实都源自某个最简单的想法”——扎哈那个所谓的“简单想法”就是“曲线”，那象征着现代生活的“疯狂”和“有机”。在上述女性看来，伊拉克裔英籍女建筑师扎哈无疑是个前卫的孤胆英雄，“哪怕她设计得再好，他们也不待见这个女人，可她还是成功了！”的确，被讥为“纸上建筑师”的扎哈屡遭歧视和挫折，曾以教课为生，43 岁才设计建成第一件作品，但她 50 岁后时来运转，以辛辛那提当代艺术中心的设计成为首获普立兹克奖的女性，从此高歌猛进，化为传奇。扎哈的性格火爆强硬、浪漫不羁，她的建筑也同样大胆张扬、流动多变，以阿利耶夫文化中心最具代表性，其形式前所未见，这既归功于她天马行空的想象力，又得力于 CAD 软件和参数化设计。从还在伦敦 AA 建筑学院跟随库哈斯学习起，扎哈就开始探索“漂浮”的建筑，而才华横溢的她在家具服饰和“超级游艇”的设计上也很出彩，由她来设计深圳浪骑游艇会真合适无比。

ABOUT THE ARCHITECT

"Architecture is rather easy," A lady who pays high respect to Zaha says, "It might look complicated structurally, yet the original concept really is simple." —That "simple concept" to Zaha is just curves, curves that implicate the crazy and organic modern lives. From that lady's point of view, Zaha, the Iraqi-British female architect, was no doubt a loner and a pioneer heroine. "No matter how good her design was, as a female she was never truly respected, but she got recognized successfully anyway." In fact, during her early career Zaha was notorious for her so-called "paper-based design" ideas, she used to teach for a living and her first project was not built until she was 43 years old. Surprisingly when she was 50, she became the first woman to receive the Pritzker Architecture Prize with Contemporary Arts Center in Cincinnati. From that point Zaha had advanced triumphantly and become a legend. She had a romantic yet short-tempered personality, which reflected in her fluid and bold designs. For example, Zaha's representational project, the Heydar Aliyev Cultural Center, had expressed a form that had never before encountered, which combined her superior imagination and advanced CAD paramatric design technology. Zaha's exploration of "floating" architecture had started back when she was still studying with Koolhaas in AA, London, her brilliant talent also lied in architecture, fashion and "super-yacht" design, which had made her the perfect candidate to propose the Shenzhen Long-Cheer Yacht Club.

扎哈·哈迪德
Zaha Hadid

1. 阿塞拜疆共和国阿里耶夫文化中心
2. 模型渲染示意图
3. 总平面图渲染图
4. 渲染图

1. Heydar Aliyev Centre
2. 3D Render
3. Masterplan Render
4. Render

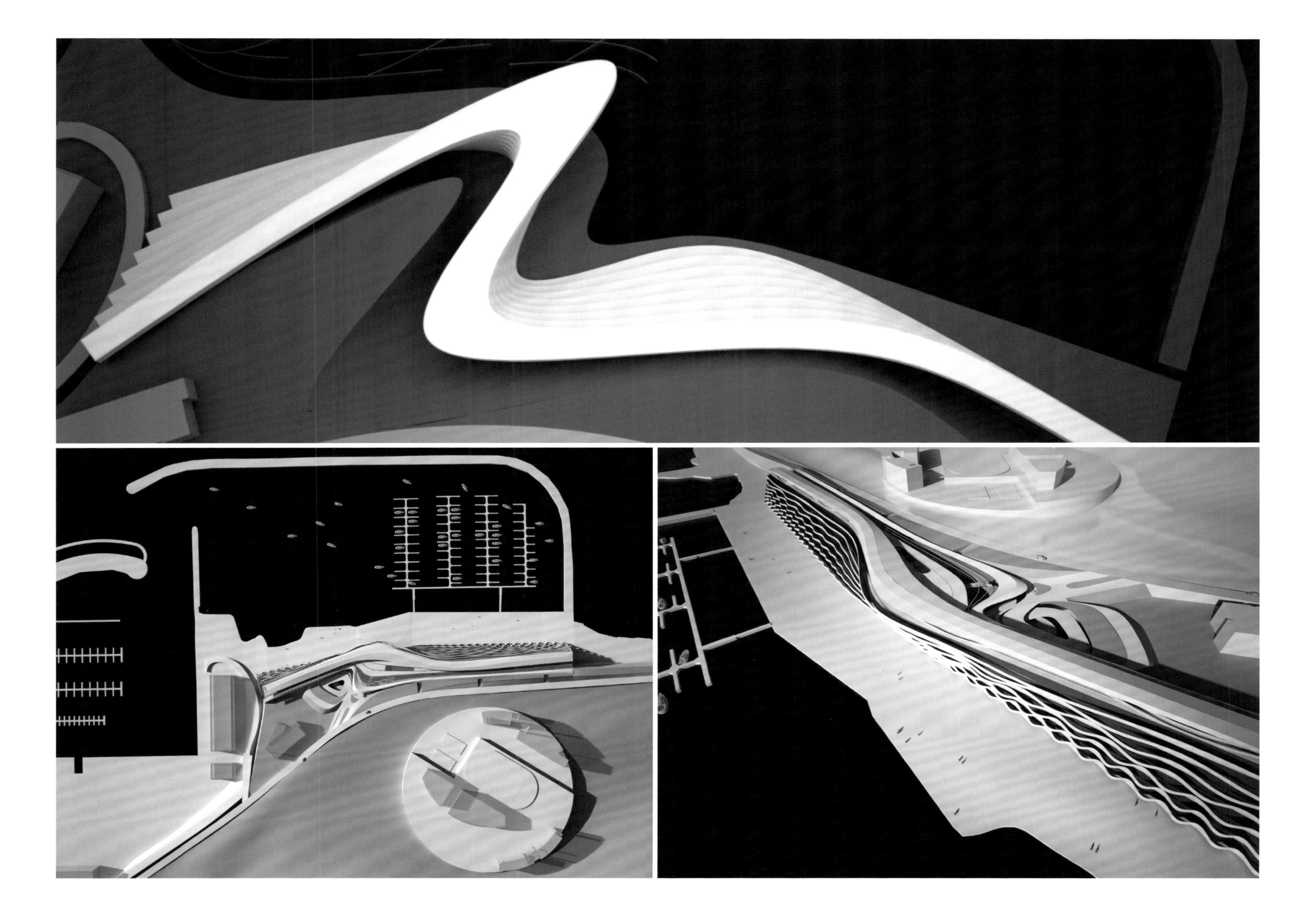

11

Mei Li Zhou Church in Liangzhu

良渚美丽洲堂

永恒与纯净的空间

建筑设计：津岛建筑事务所

项目地点：杭州市余杭区万科良渚文化村

设计时间：2010年

建筑总面积：1,025㎡

功能：教堂，图书馆，社区中心

Arch. design: Tsushima Design Studio

Location: Vanke Liangzhu Cultural Village, Yuhang District, Hangzhou

Design process: 2010

Total area: 1,025m^2

Program: Church,Library, Community Center

美丽洲堂坐落在杭州一处树木繁茂的地方，其设计不仅仅融入环境，也很好地为社区生活服务。美丽洲堂在为宗教服务的同时，也为周边社区服务。教堂力求融入自然的环境当中，不仅重视功能和用户体验，还选择了低碳材料，最大限度地降低了对环境的影响。同时建筑师做了广泛的研究，让景观与建筑边界模糊，使得建筑与景观无缝接入。

美丽洲堂的设计有两个重点，第一个是室内空间与自然之间宛然一体，第二个是表达永恒和纯净。

美丽洲堂由三个单体建筑（美丽洲教堂、小花园教堂、办公处）和一个钟楼组成。其中两栋建筑较小，每栋材料各异但是整体统一。美丽洲堂是与自然的门户无缝衔接的大型开放式建筑。简单的平面，广阔的内部空间，光线和自然流动其中，大型垂直天窗坐落在两端上方。与美丽洲堂的高大相比，小花园教堂和办公区建筑较小，落地窗传递出与自然更好的联系。虽然它们看起来是大教堂的演变产物，其实各有千秋。庭院联系整个建筑群，每一处的景观都独一无二。在景观序列体验上，给人们更丰富的感受。教堂完美地融入了周边森林，同时外观混凝土与内部木材的对比体现了轻与重的反差。未做雕琢的景观石材将每块石材本身的特性优雅地表达了出来。

我们在探讨永恒和纯洁意义的同时，也让建筑结构与四季风貌相协调，以给予人心灵的震撼。建筑师同样注重细节，木结构去掉了所有的装饰要素，非常简洁和庄严，同中国建筑相比，这个建筑显得更为日系。屋顶空间是一个精致的序列化空间（11m x 35m）。木结构之外，墙壁也是未装饰木材，地板是白色的瓷砖。木材，光，开放的室内空间，以及四季变化的景观，共同谱写永恒之曲。

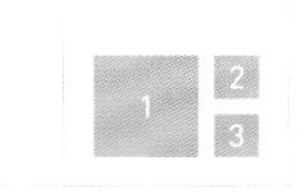

1. 总平面图
2. 平面图
3. 剖面图

1. Masterplan
2. Plan
3. Section

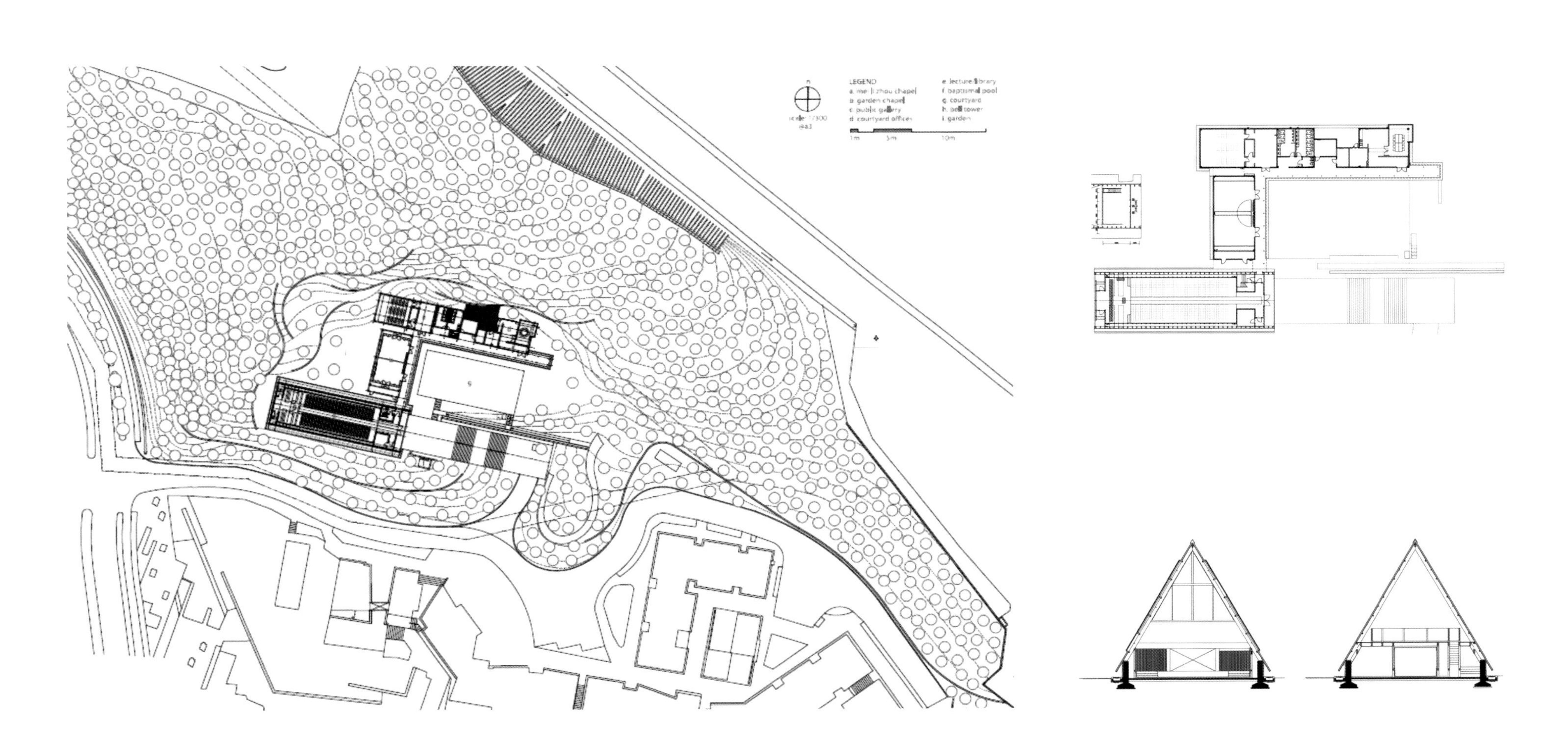

Nestled within a wooded development in Hangzhou, China, the Mei Li Zhou Church was an attempt to seamlessly merge not only into the existing natural environment, but also into the lives of those who live within its surrounding community. Carefully developed in coloration with TDS (Tsushima Design Studio), and our client Zhejiang Vanke, Mei Li Zhou Church was not only created for religious services, but as both a spiritual and community retreat for all of the surrounding community regardless of religious beliefs. Surrounded by nature, the Mei Li Zhou Church sits not as an object to be viewed from afar, but rather attempts to blend in naturally into the surrounding environment. During the design process not only was user experience and functionality of great importance, but also working with environmentally low impact materials and minimizing environmental damage were both very important to the projects success. As a result of our extensive studies in architecture and in landscape, the churches boundary blurs with the surrounding landscape, allowing for a seamless flow between building and nature.

The Mei Li Zhou Church was designed with two overwhelming themes. First the creation of a seamless flow of space between the interior spaces and nature, while the second looks at the idea of timelessness and purity.

The church complex consists of three separate, yet connected buildings (the Mei Li Zhou Chapel, the smaller Garden Chapel, as well the Courtyard Offices) and a single bell tower. The two smaller buildings are considered public galleries. Although each building utilizes different materials and design features, the concept of flowing spaces connects them together. The Mei Li Zhou Chapel serves as the gateway to the project in both architecture and nature. Its simple plan, vast interior space and large open facades allow for the seamless flow of nature and natural lighting into the building. Along with the two large openings at the front and rear of the chapel, a series of large vertical skylights are located over the pulpit creating a very ephemeral feeling within the church.

While the Mei Li Zhou Chapel utilizes its height and large views to bring nature in, the two smaller buildings use very carefully placed floor level windows to allow the spaces to flow seamlessly. Although both the main chapel and the smaller buildings evolved from the same concept, the resulting design and created views are quite unique. The buildings integration with the landscape allow for the creation of a seamless flow of space throughout the project. Situated in the center of all three buildings and housing the Bell Tower is a large outdoor public courtyard. The courtyard acts as a device, which continues through each of the three buildings, connecting them to each other as well as nature. In the two gallery buildings, each room is carefully connected to this courtyard creating a series of interconnected spaces. Each space is unique and offers different views of the surrounding landscape. This sequence of flowing space guides people on a path, encouraging them to explore the buildings and all of the experiences they have to offer.

On the other hand we examined the concepts of timelessness / purity, and what that means for the architecture and its inhabitants. How can a fixed object, within an ever-changing environment blend seamlessly from the changing colors of the seasons, to the changing of people's beliefs and cultures? For us this timelessness was found in the materials and details. This timelessness can be found in the main chapel, and its wood structure. Stripped of all decorative elements, this building stands pure and dignified. In a time when concrete structures are norm in Chinese construction, our structure looks to a more traditional Japanese wood system. Designed and constructed in Japan, the engineered wood roof allows for a beautifully detailed column-less space of 11m x 35m. Besides the wood structure, the walls are clad in unfinished wood, in contrast to the white tile flooring. Timelessness is created in the colors of the wood, light, and openness of the interior spaces which are constantly changing in conjunction with the changing of nature surrounding the building.

1. 美丽洲堂室内

1. Interior of the Church

“只愿做自己愿意住的建筑。”

关于建筑师

“城市的过去与未来是有连续性的”，津岛晓生在留美游欧后感悟到，“未来是由过去建成”。自从他在东京创立津岛设计事务所后不久接手上海万科中心的项目起，津岛就开始研究中国建筑史，思考用传统材料塑造能够融入周围景观的现代建筑。在杭州这座他眼里纯粹而自然的城市，津岛将良渚文化村的美丽洲堂设计成简洁的纯木结构建筑，使用山木、木纹板和当地的石、瓦等材料来体现自然生命力——他不仅保留了原地的树木，还在清水混凝土台面上加了木纹，平淡精致，别有生趣。外国教堂往往神圣肃穆，给人以距离感，但在美丽洲堂里，窗一开就是远山树影，颇有东方文化中天人合一之意。津岛每次回杭州都会去看看美丽洲堂和周围树木，看看是否让“这座建筑能够自然生长，最终与环境融为一体”。这种建筑与环境交融共生的“可触空间”美学被其事务所奉为圭臬，在近年的镰仓别墅、苏州新纽顿会心街国际幼儿园、北京随园等设计中都有体现。

ABOUT THE ARCHITECT

"There is a continuation between the past and the future of a city," Toshio Tsushima signed after his studies in the US and travel in the Europe, "The future," he pointed out, "is built by the past." Soon after his establishment of Tsushima Design Studio (TDS) in Tokyo, Tsushima had taken over the project of Shanghai Vanke Center. He began to research Chinese architecture history, and aimed to design buildings with modern concept that uses traditional materials to mingle with the surrounding environment. Hangzhou, a city of purity and nature, had greatly inspired Tsushima in designing the Mei li zhou Church in the cultural village. The church is built with simple timber structure, combining unfinished woods, wood-grained boards and local stones and tiles, it is abundant with nature power. The existing trees are completely preserved, and ever the concrete surfaces are engraved with timber finish. When we think of church in the western culture, there always seem to be a solemn holy place with certain feeling of distance. Here at Mei li zhou Church, the distant mountain and shadows of trees had provided an intimate eastern atmosphere, where human and nature are integrated closely. Every time when Tsushima travels to Hangzhou, he would visit the Mei li zhou Church and check on the trees, to see if this building could "grow with the nature and merge back into the environment eventually". Such "Tangible Space" concept is the foundation of TDS design, their recent projects, Kamakura Villa, Suzhou NEWTON Kindergarden Huixin Campus and Beijing Suiyuan Project present deep understanding of the correlation between architecture and nature.

津岛晓生
Toshio Tsushima

1. 天竺市场中心
2. 美丽洲堂内部木结构
3. 外景图

2. Tianzhu Market Center
3. Timber Structure Inside the Church
4. Exterior of Church

1 2

1. 建筑
2. 室内

1. Exterior of the Church
2. Interior of the Chaple

12 Liangzhu Culture Center

良渚文化艺术中心

安得广厦万间，大屋顶下聚欢颜

建筑设计：安藤忠雄建筑研究所
项目地点：杭州市万科良渚文化村
建筑总面积：12,812.9 ㎡
功能：文化艺术中心

Arch. design: Tadao Ando Institute of Architecture
Location: Vanke Liangzhu Cultural Village, Hangzhou
Total area: 12,812.9m^2
Program: Culture Center

日本建筑大师安藤忠雄设计的良渚文化艺术中心位于中国杭州良渚文化村，这是一个结合生态、观光与人文艺术的大型住宅村落计划。建筑室内分为展示栋、教育栋及有图书馆的文化栋，一并被覆盖在巨大屋顶下，这也是中心被村民昵称为“大屋顶”的由来。这位具有传奇性的现代建筑设计大师，凭借独特的清水混凝土应用，抽象化的光、水、风表达，引发风潮。而良渚文化艺术中心将安藤忠雄脑海中对建筑的这些美好设想一一呈现。

主体运用清水模工法，呈现混凝土自然朴实、不加修饰的刚硬质感；在结构单纯、看似封闭的内部设计了数十个三角形采光窗，引入天光，随时间推移产生丰富的光影变化；建筑周围种植了整片樱花林，有图书馆的东侧围绕着浅水，夜晚来临时水面倒映出直线窗框与室内光影，和巨大展示书格营造出几何感强烈、壮阔而富有禅意的空间。

Located in the historic Liangzhu Village in Hangzhou, a new monumental cultural center was designed by the Japanese architect Tadao Ando. The building is another popular cultural site in the village as part of the grand-scale residential village masterplan project that focuses on ecology, tourism and humanitarian-art. The interior of the building contains three blocks: Exhibition Hall, Education Building and a Cultural Hall consists a library. The three blocks are covered with a giant roof-structure, hence the nickname of the building by the local villagers. Tadao Ando, a legendary modern architecture designer, continued his iconic use of concrete and strikingly abstract representation of light, water and wind, had captured these concepts in this marvelous design.

The major hall is constructed with clean-water molded concrete, exhaling the raw and natural feature of the material, giving a sharp finish. The seemingly-simple enclosed interior is lit by dozens of triangular-shaped skylight, the play with natural light and shadows alters at different times in a day. The building is surrounded by a tranquil sakura landscape, shallow water surface hugs the eastern side of the library, creating a strong visual impact at night with the contrast between the vast exhibition book shelf and its reflection in the water. The ever-changing light and shadow inside the building, together with the straight-lined window frames, has made the architecture a chamber of spiritual zen tranquility.

1. 手绘草图
2—3. 平面图

1. Sketch Design
2—3. Plan

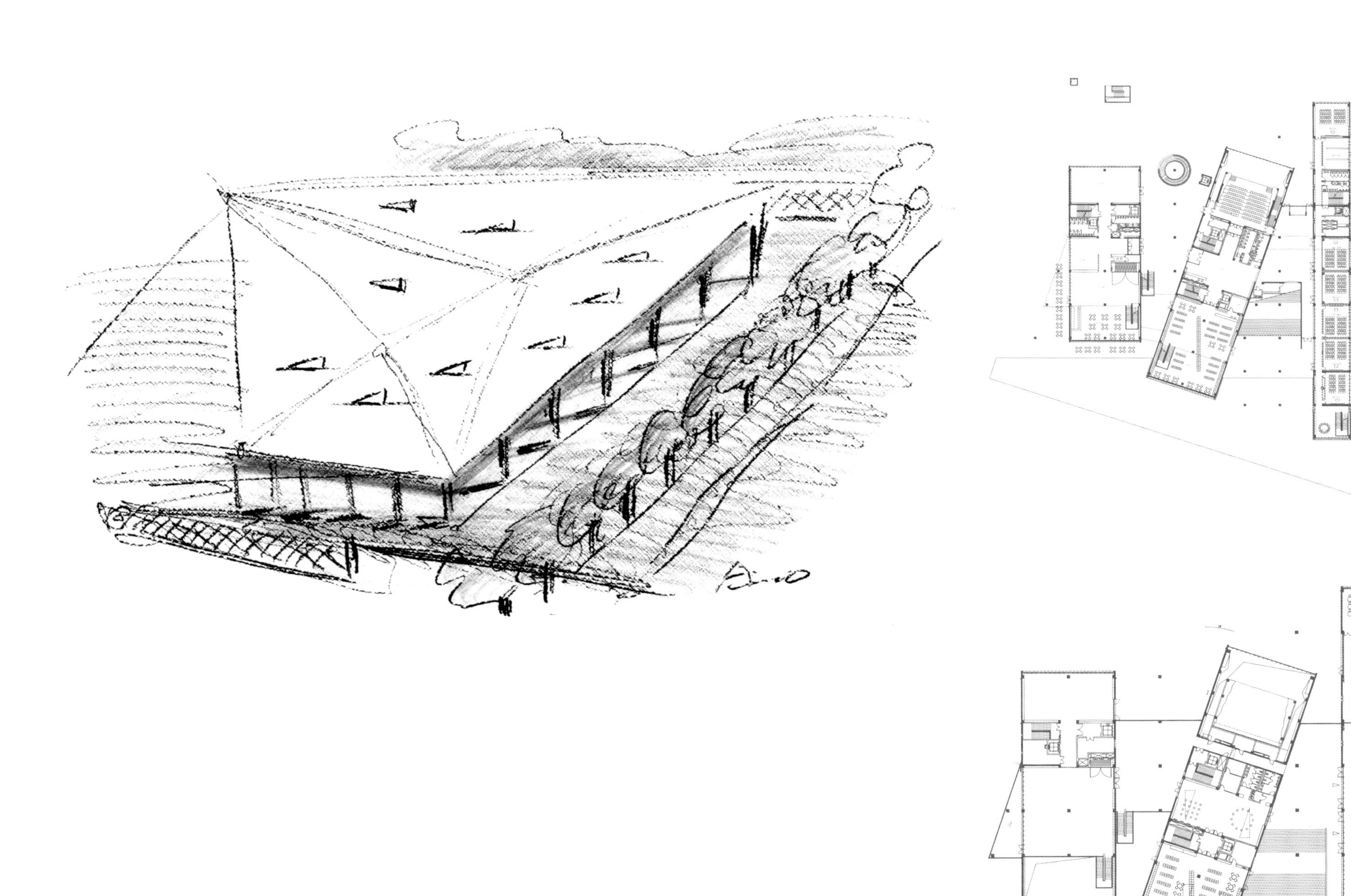

“如何将留存下来的历史印记与未来的时代相结合，这是我们建筑师的重要使命。”

关于建筑师

ABOUT THE ARCHITECT

“我的人生经历中，找不到可以称为卓越的艺术资质，只有与生俱来的面对严酷现实，绝不放弃，坚强活下去的韧性。”——这句话或许普通，但出自自学成才的传奇建筑师安藤忠雄之口，就有了非凡的意义。这位在日本家喻户晓的励志楷模高中时就成了战绩不错的职业拳击手，18 岁立志做建筑师的他一边打工一边竭力了解他想知道的一切，靠积蓄环游日本并去世界各地旅行，28 岁时回大阪创立了安藤忠雄建筑研究所。作为没上过大学的草根，安藤忠雄将拳击手的意志和坚韧用在了进行“建筑游击战”的创业之路上，以“住吉长屋”等作品获得业界肯定，确立了以清水混凝土和几何形状为标志的个人风格，又以“光之教堂”等神来之笔成就了一代经典，荣获 1995 年的普利兹克奖并声名显赫。已对抗癌症十余年的安藤忠雄仍在奋斗中获得幸福，因为“一个人真正的幸福并不是待在光明之中。从远处凝望光明，朝它奋力奔去，就在那拼命忘我的时间里，才有人生真正的充实”。

"Throughout my experience of life, I have not discovered what may be called outstanding artistic qualifications. Rather, only the inherent ability to face the harsh reality, to never give up, and maintain a strong resilience to live." This sentence may seem common, but is borne from the mouth of the self-taught legendary architect Tadao Ando, and it holds extraordinary significance. The well-known inspirational model in Japan became an impressive professional boxer in high school. At the age of 18, aspiring to be an architect, he worked hard to understand everything he wanted to know. Relying on his savings to travel around Japan and the world, he returned to Osaka at the age of 28 to establish the Tadao Ando Institute of Architecture. Having never attended university, Ando relied on his boxer's will and perseverance on his creative journey of "architectural guerrilla warfare". His works such as "Azuma House" garnered high recognition within the industry and established his personal style marked by clear concrete and geometric shapes. He found great success in works such as the "Church of Light", a stroke of genius which became a generational classic, and which earned him the 1995 Pritzker Architecture Prize. Ando, who has been fighting cancer for more than ten years, is still struggling towards happiness, because "a person's real happiness is not found in the light. Staring at the light from a distance and striving towards it, only in that time of desperate self-forgetting may life be truly enriched".

安藤忠雄
Tadao Ando

1. 光之教堂
2. 博物馆鸟瞰图

1. Church of Light
2. Birdeye View of the Museum

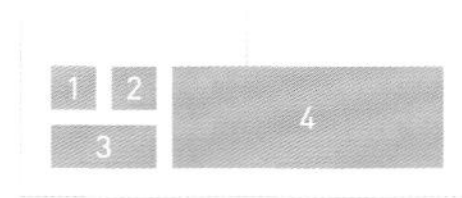

1–3. 外景
4. 室内

1–3. Exterior views
4. Interior views

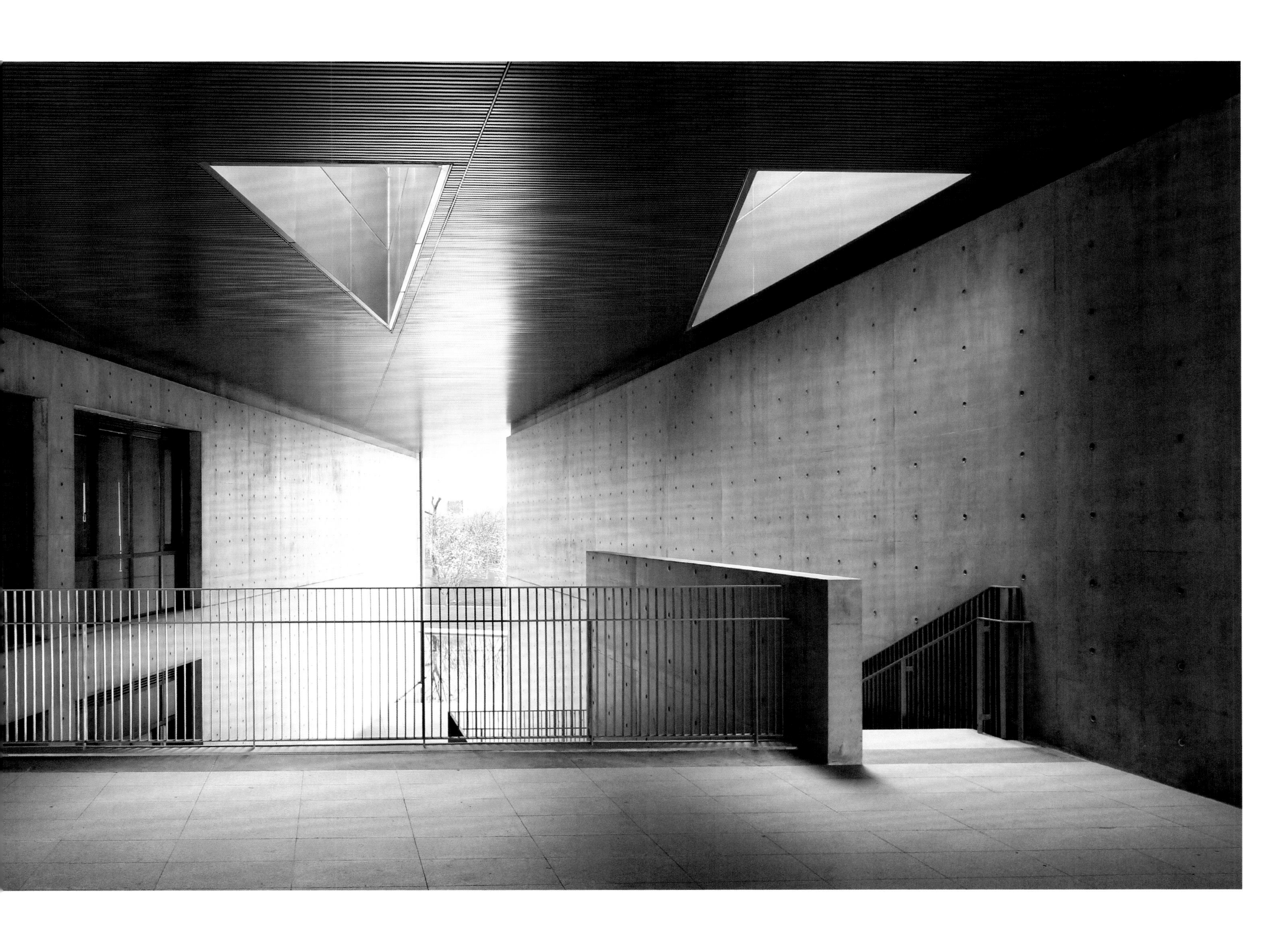

13 Deep Dive Rowing Club

深潜赛艇俱乐部

穿梭在水林和心灵之间的小艇

建筑设计：山水秀

项目地点：上海市浦东新区世纪公园

设计时间：2016 年

完成时间：2017 年

建筑总面积：300 ㎡

功能：青少年赛艇培训和交流活动

Arch. design: Scenic Architecture Office

Location:Century Park, Shanghai Pudong District

Design process: 2016

Completion time: 2017

Total area: 300m²

Program: Rowing Training and Activities for Teenagers, Pier, Boathouse, Lockers & Showers, Activity Room

为了长期推广和助力中国的赛艇运动，万科教育集团与上海浦东新区合作，计划在世纪公园内建造一座小型青少年赛艇俱乐部，为更多的青少年组织赛艇培训和交流活动，计划每年学员约 200 人，每期 20—30 人。俱乐部需要存放约 15—20 艘赛艇，并为青少年提供培训活动室、淋浴间、洗手间和休息区。俱乐部对学员进行日常培训和技术指导，同时也对学员家长和赛艇运动爱好者开放。

世纪公园是上海市中心城区内对公众开放的最大的湿地公园。可以进行赛艇培训的张家浜河宽约 35 米，是一条横穿公园的城市内河。赛艇运动员是背对前进方向划桨的，出于对安全和弯道变速的考量，山水秀的方案将俱乐部设在河道弯折处一个附带小港湾的地点，便于双向出发和抵达。基地上是密植的水杉林，仅在岸边留有一座 4 米宽的老码头。为了最大限度地减少对原环境的影响，使用空间被分成四个部分：浮式码头设在南侧的港湾里，活动室放在西侧的河道中，更衣室则建在老码头原址上，艇库是项目中唯一需要占用林地的部分，为了避免大面积砍伐或移栽，艇库拆分成三个细条散落在水杉林里，大致对应着码头的走向。

赛艇分单人、双人、四人和八人四种，长度在 8—18m，宽度只有 30—60cm。为赛艇设计的“小家”宽度以容纳一条艇为限，上下共 4 层，长度按四种艇的长度模数组合而成。为了减少现场施工对公园的影响，艇库的结构基础由预制的点状混凝土块构成，用两片角钢栓接起来的柱子如小树干一般，在上部分叉成 Y 形，支撑顶部的人字形钢板雨棚，在下部则单侧悬挑出四层“枝干”，用来放置赛艇。纤细的艇库除必要的顶部遮蔽外完全敞开，让取还赛艇成为一种负重在肩的林中漫步。为了减少移栽，基地内每一株水杉的位置均被测量和记录，让三条艇库可以从合适的角度插入林中。沿着树影，现场放样找到了合适的空地和角度。

更衣室建在原有码头的位置上，是一座窄条形的房子，四周用巴劳木板墙围护，屋顶是由山形梁支撑的钢折板金属屋面，通过顶部天窗给更衣和浴室空间带来自然光。建筑师希望这座小屋的实体性既能满足自身对私密的需要，也能在艇库栖身的杉林和活动室置身的河流之间形成一个屏障，围护这两个场所各自的单纯体验。

水中的活动室是一座类似“不系之舟”的水榭。建筑底面是一个类似驳船的长方形钢格板，由打入河床的管桩支撑，覆盖空间的是一个 20m 长的双坡屋盖，仅由位于两端的 H 型组合钢柱支撑——这为空间的营造提供了自由和因借：H 形双柱之间是通往亲水平台的门洞，柱跨上的双梁之间则成为顶部采光的通道。活动室东侧紧邻更衣室，一条 8m 长的柜子将两个空间划分开来，前为储物，后为书写板，两端则容纳了空调柜机。活动室的其他三面朝向开敞的河景，临水的西侧特别采用了三幅折叠推拉窗扇，可以对室外完全开放；推拉扇之下是可以安坐的通长窗台，使这一内外空间的边界成了休憩、交流和观景的场所，也在室内的测功仪培训和河里的赛艇训练之间建立了视觉联系。虽然整座“水榭”在空间上与四周内外流通，但又通过其水中的方位以及比河岸略低的标高，赋予它一种特殊的场所感。从更衣室端部的入口需要下三级踏步才能进入这个船舱一般的空间，而置身其中，又能在安定之余感受到与外界自然的通达融合。

赛艇码头位于基地南侧河道与港湾的分界处，这座浮式码头通过抱桩滑轮与水中的管桩固定，上面满铺塑木板，由捆扎在一起的浮筒群承托。码头两侧都可以停靠赛艇，东侧通过一个 5.5m 宽的坡道与岸边连接，通往杉林艇库；西侧则通过一部小梯与活动室和更衣室之间的通道联系起来。坡道和小梯的两头都采用了铰接节点，以顺应水位的涨落。

码头、艇库、更衣室、活动室是俱乐部需要的四个独立场所，一组便捷并且合乎逻辑的动线把它们有机地联系在一起。从公园道路旁顺着一条小径进入杉林，在接近俱乐部入口时会遇到一个分岔，向左继续穿行在林间的是运送赛艇的通道，向右则可穿过杉林，抵达入口处由更衣室、木板墙和一株大柳树围合而成的半开放庭院。左边的路径在杉林里再次分散到各条艇库，而后又在岸边再次聚拢，通过坡道抵达码头；右边的路径进入室内，成为更衣室和活动室之间的通道，然后再穿出至室外岸边，通过小梯抵达码头；至此，两条路径又在岸边和码头得以汇合。

在这个动线系统中，户外部分除了高大的水杉林，还有丰富的灌木、花草植被，以及经常在林间和河畔活动的松鼠、乌龟等小动物。为了减少对它们的打扰，仅在通往建筑入口的路径上使用了木板铺装，在艇库区域则采用了通透的设计。沿着运艇的路线，600 块点状的小混凝土块作为基础，上面放置不锈钢金属格栅作为步道。这样的通透式步道仍然允许植物在其间生长，小动物的活动也不会被道路打断。

希望这座俱乐部不仅能助力青少年赛艇运动的推广和发展，还能够通过建筑与自然环境的友好相处，对青少年生态观念的养成起到潜移默化的影响。

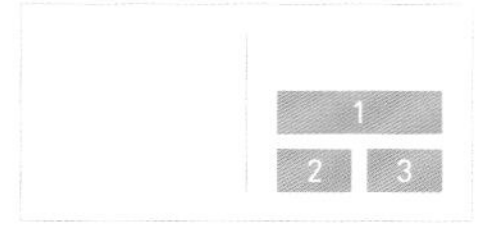

1. 总平面图
2. 屋顶平面图
3. 前期分析图

1. Masterplan
2. Roof Plan
3. Analytical Diagrams

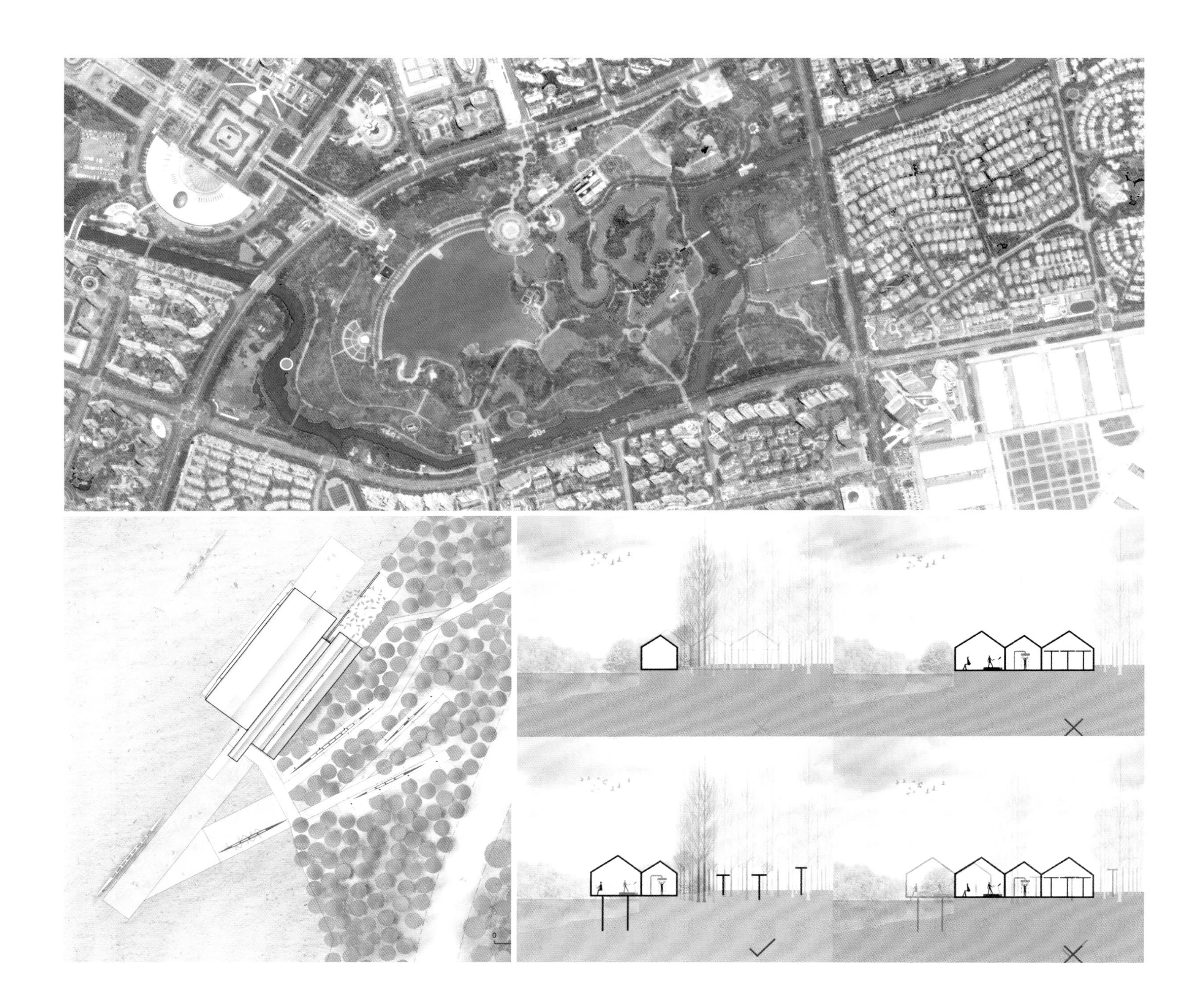

In order to promote and support China's rowing sport for a long term, Vanke Education Group, in collaboration with Shanghai Pudong New District, planned to build a small rowing club in Century Park to organize rowing training and activities for more youth and teenagers. Having 200 trainees each year with 20 to 30 every month, the club needs to store approximately 15-20 rowing boats and provide training & activity rooms, lockers & showers, toilets and seating areas for adolescents. The club will conduct daily training and technical instruction and also open to parents and rowing enthusiasts.

Century Park is the largest wetland park open to the public in downtown Shanghai. The Zhangjiabang River, which can be partially used for rowing training, is an urban river that crosses the park in a rough width of 35m. Rowing athletes face the opposite direction of advancing when paddling. Due to the considerations of safety and speed change, the scheme set up the club at the bend point of the river with a cove to facilitate two-way departures and arrivals. The site was covered with densely planted metasequoia forest, leaving only a 4-meter-wide old pier on the shore. In order to minimize the impact on the original environment, usable spaces are devided into four parts: the pier is located on the south side of the cove, the activity room is placed on the west side of the river, and the changing room is built on the old pier site. The boathouse is then the only part of the project that needs to occupy the existing forest. To avoid large-scale felling or transplanting, the boathouse is split into three thin strips scattering in the forest, roughly corresponding to the direction of the new pier.

The rowing boats have 4 kinds in size for singles, doubles, fours and eights. The length is about 8-18m and the width is only 30-60cm. The width of the "small home" for the boats is limited to accommodate only one boat yet with four floors, and the length is a combination of the boat length modules. In order to reduce the impact of on-site construction in the park, the structural foundations of the boathouses consist of prefabricated point-like concrete blocks. The column composed by two bolted angle steel, is forked in the upper part like a tree to form a "Y" to support double pitched canopy. On the lower part four "branches" cantilevered from the column serve the rowing boat storage. The slender boathouses are completely open with only necessary roof shelter, making the boat carrying a weight-bearing walk in the forest.

To reduce transplanting, the position of each Metasequoia in the site is measured and recorded so that the three boathouses could be inserted into the forest from suitable angles. However, it is still difficult to position them on site. Along the shadows the team adjusted the angles and find more spaces.

The changing room built on the original pier is a narrow bar enclosed by Bangkirai panels. The roof is made by folded steel plates supported by gable-shape beams, with skylights bringing natural light into lockers and showers. The solidity of the changing room will not only satisfy its own need for privacy, but also help us as a barrier between the forest for boathouses and the river where the activity room was located, hence to enclose the pure experiences of these two places.

The activity room in the river is a waterfront pavilion like "a boat not tied". The bottom is barge-like rectangular steel grating on pipe piles driven into the riverbed. The cover is a 20-meter-long double-pitched roof, supported only by twin steel H-pillars at both ends this gives freedom and access to the space: between the H-pillars is the door towards exterior platforms, between the twin beams on the pillars becomes the passage for the skylight. The east side of the activity room is adjacent to the changing room. An eight-meter-long cabinet divides these two spaces, the front is for storage, the rear is a writing board, and the two ends contain the air-conditioners. The other three sides of the activity room are facing the riverfront; on west side three sets of foldable sliding windows provide complete openness to the outside. Below the sliding windows is a long and wide wooden sill that can be seated, making this boundary space a place for rest, communication, and viewing. It also established a visual connection between indoor ergometer training and rowing training in the river. Not only the pavilion is allowed to open itself to the surroundings, but also give it a special identity through its position in the water and a slightly lower floor elevation in comparison with the riverbank. To enter the pavilion requires three steps down from the entrance, it is yet until placing yourself inside this calm space that you start to perceive the merge with the nature outside.

The pier is located at the boundary between the river and the cove on the south side of the base. The pier floats in the water and is fixed with pipe piles by embracing pulleys. It is covered with plastic wood panels and supported by bundles of buoys. Serving both sides, the pier connects to the shore through a 5.5-meter-wide ramp to the boathouse area, and to the passage between the activity room and the changing room through a small ladder. Both the ramp and the ladder are connected by hinges to comply with the fluctuation of the water level.

Pier, boathouses, changing room, and activity room are four independent venues that the club requires. A set of convenient and logical circulation that can link them in an organic way. Entering from the park road, following

1. 剖面图
2. 为植物、动物和使用者着想的友好设计
3. 构造剖面—活动室

1. Sectional View
2. A User-Friendly Design for Both Human and Animals
3. Activity Room Detail Section

a path into the forest, you will encounter a fork near the entrance to the club. Turn left you will continue walking through the forest on the lane for transporting the boats. Turn right you can cross the forest and arrive at the entrance, which is a courtyard semi-enclosed by the changing room, the extension of a wooden panel wall and a large willow tree. The left path was dispersed to the boathouses in the forest, then gathered again on the shore and reached the pier through the ramp; the path on the right enters the building and become a passage between the changing room and the activity space. When passing out, it arrived pier as well through the ladder. At this point, these two paths were merged again on the pier.

Apart from the tall metasequoia trees, in this circulation system the outdoor area has abundant shrubs, flowers, vegetation, and small animals such as squirrels and turtles that often move between forests and riversides. In order to reduce interfering their environment, wood plank paving is used only on the way to the building entry, but make permeable paths in the area of the boathouses. Along the paths there are 600 small concrete blocks as individual dots, on which stainless steel grills were placed as trails. These transparent walkways still allow plants to grow between them, and keep rooms for the small animals.

This club is designed not only will help the promotion and development of the rowing sport, but also be able to exert a subtle influence on the establishment of teenagers' ecological idea through a friendly relation between architecture and natural environment.

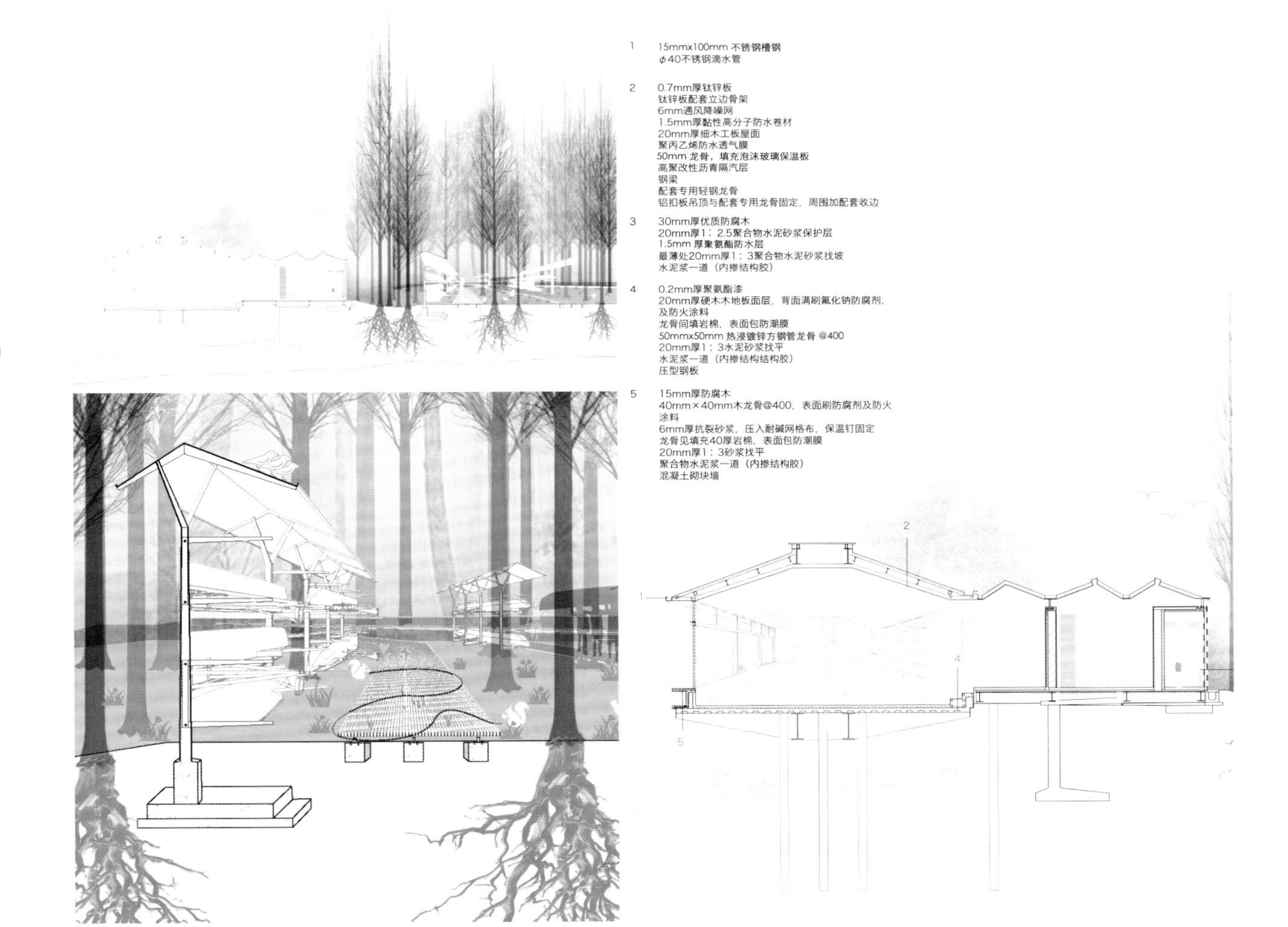

关于建筑师

山水秀（祝晓峰、杜洁、周延）

祝晓峰小时候爱画山水画，还梦想“在有山有水的环境里盖自己的房子”，于是他 2004 年回到上海创办建筑事务所时就想到了“山水秀”这个名字。他的山水情结使其事务所关注建筑对传统的继承以及与自然的融合，然而中国尽管在过去三十年里飞速发展，但在文化层面却有着明显的断裂。祝晓峰自己也不例外，坦言“曾经对上世纪中叶的中国建筑师了解不够”，至今仍在“补课”，学习陈其宽、王大闳等人在延续传统和突破界限方面做的努力，积极探索“发自传统本身的现代性”和本土化的原创建筑语言。对这位曾在纽约 KPF 建筑事务所工作五年的哈佛硕士来说，他回国创业的动力就是要“实现自己理想的工作状态”，在项目选择性和实施完成度有一定保障的情况下促进传统文化的创造性转换。朱家角人文艺术馆就是其代表作之一，有传承，有突破，还有别致的体验感。在祝晓峰看来，中国建筑师一方面要融入全球化进程，另一方面也要带动中国文化的复兴。

ABOUT THE ARCHITECT

Scenic Architecture (Zhu Xiaofeng, Du Jie, Zhou Yan)

Since his childhood, Zhu Xiaofeng has become a big fan of Chinese Shan-Shui paintings, and wished to build his own house that is "surrounded by mountain and water".In 2004 he returned to Shanghai and set up his design studio called "Scenic Architecture", and started to fulfill his Shan-Shui complex in his designs that focus on carrying on traditions and relations with nature. He pointed out that there had been a severe cultural dislocation in the past 30 years of the fast-development of Chinese reform process. "I don't have much understanding of Chinese architects in the middle of last century, " Xiaofeng admits, "and I am constantly learning." He had looked deep into the works by Chen Qikuan and Wang Dahong, of which they had seek methods in inheriting traditions while crossing boundaries, and tried to develop a vocabulary system for original local architecture design that "express a modernity within his authenticity". Zhu Xiaofeng had worked for KPF for five years in New York after his Harvard master studies, he returned to China to "work under an ideal condition", which means to facilitate the creative transformation of traditional culture under carefully selected projects that would ensure a relative complete execution level. Zhujiajiao Museum of Fine Arts was one of its representational works with tradition preservation, pioneer design concept and unique visiting experience. From Zhu Xiaofeng's perspective, as a Chinese architect one should participate in the globalization process while maintain the mission of domestic cultural revitalization.

祝晓峰
Zhu Xiaofeng

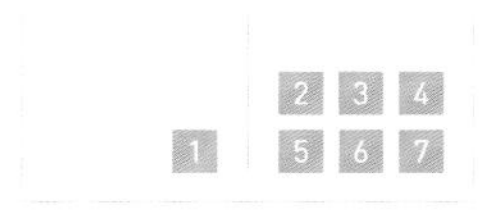

1. 华鑫中心
2. 位于世纪公园中的赛艇俱乐部，远处是上海陆家嘴中央商务区
3. 老柳树下的水上平台
4. 入口雨棚
5. 河上俱乐部和赛艇运动
6. 运艇回艇库
7. 建筑和碳纤维材质赛艇有着形态上的关联

1. Huaxin Business Center
2. Location of the Club is at Century Park
3. Floating Platform under the Old Willow Tree
4. Entry Canopy
5. Water Sports and Rowing Club Activities
6. Carrying the Rowing Back to Their "home"
7. Form Connection between the Architecture and Carbon-fiber Rowing Boats

1 2

1. 插入森林中的纤细艇库对自然完全开放
2. 森林和艇库生成的多重空间层次

1. The Slender Boathouses Are Completely Open to the Nature Environment
2. Layers of Space Between the Boathouse and the Forest

14

Liangzhu Museum

良渚博物馆

水乡泽国，又见精美历史宝盒

建筑设计：大卫·奇普菲尔德建筑设计事务所

项目地点：杭州市余杭区

设计时间：2003 年

完成时间：2007 年

建筑总面积：9,500 ㎡

功能：博物馆

Arch. design: David Chipperfield Architects

Location: Yuhang District, Hangzhou

Design process: 2003

Completion time: 2007

Total area: 9,500m²

Program: Museum

良渚博物院位于浙江省杭州市余杭区良渚街道美丽洲公园内，是良渚文化村的北段标志建筑，坐落在良渚考古场地之上，是一座收藏、研究、展示和宣传良渚文化的考古遗址博物馆。博物馆内主要陈列了良渚地区古玉文化考古发现，由英国著名建筑设计师大卫•奇普菲尔德设计，建筑非常简洁，突破了具象形态的束缚。它结合了周边的湖景，用桥梁连接了旁边的公园，并在内部建造了人工水面，巧妙地将内外的景观连为一体，体现了艺术与自然、历史与现代的和谐融合。

The museum is built on the site where many of the Liangzhu treasure were found, which is on the archaeological finding site in Meilizhou Park in Liangzhu Street, Yuhang District in Hangzhou, Zhejiang Province. The building forms the northern landmark point of the Liangzhu Cultural Village. The museum mainly houses a collection of archaeological findings from Liangzhu culture (better known as the Jade culture), designed by renowned British architecture designer David Chipperfield, the building is featured with simplified form, emerging from the lake in parts with bridges, and mingles artificial lake surfaces within the building. The terranced gardens mediate between the edifice and the water, connecting the inside and the outside, harmonically blending in art, culture, history and modern aesthetic.

1. 总平面图
2—3. 平面图

1. Masterplan
2—3. Floor Plans

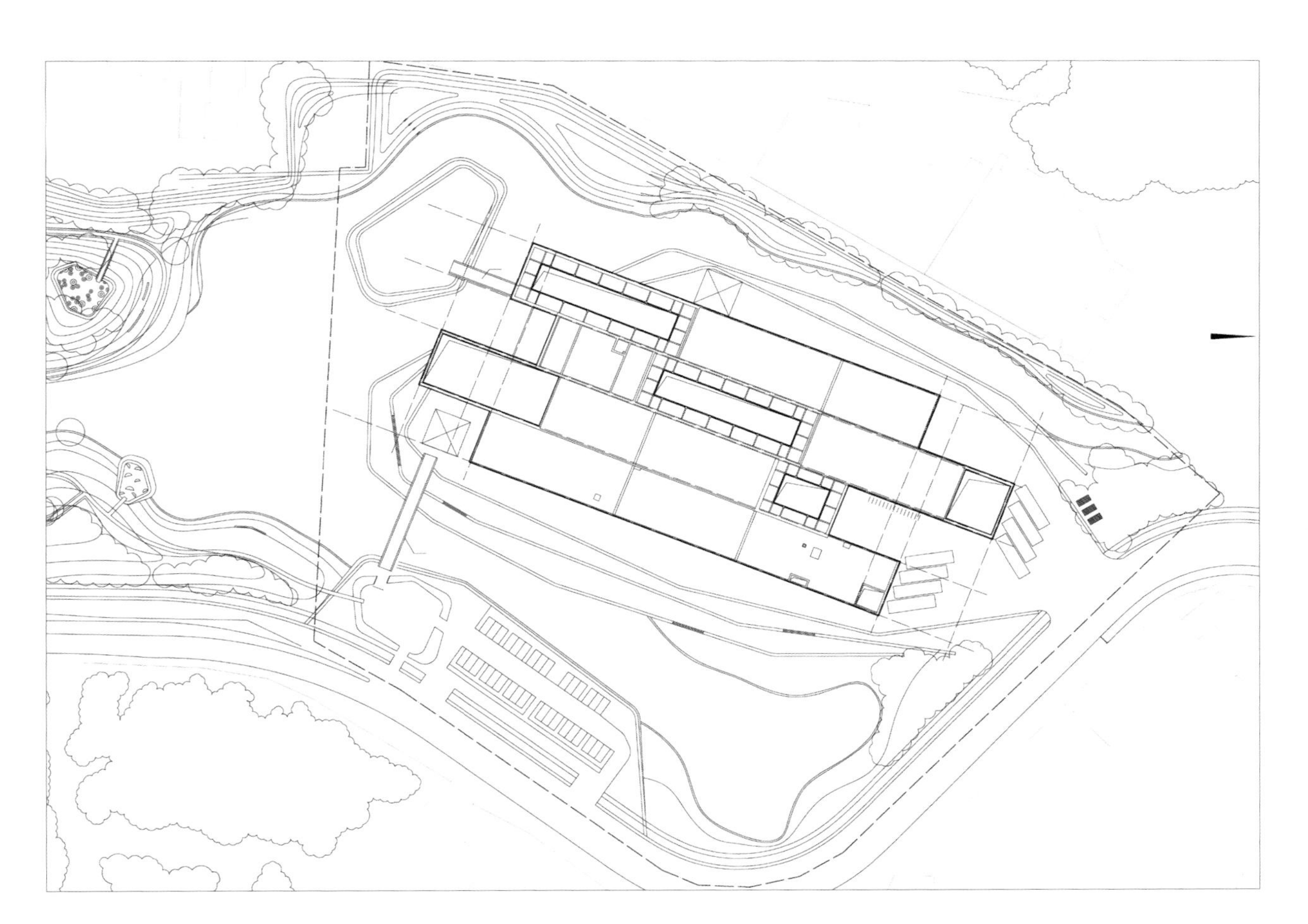

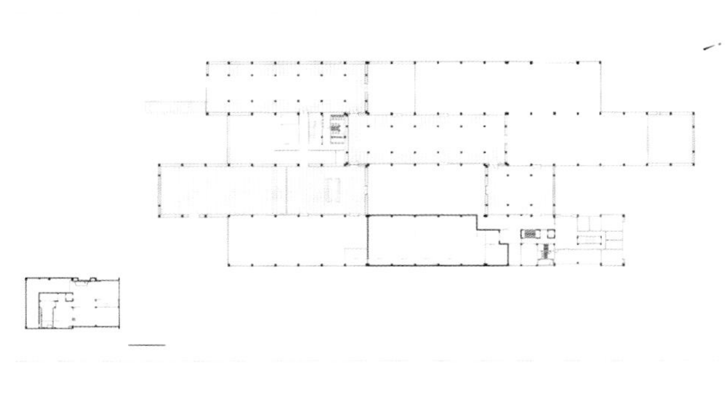

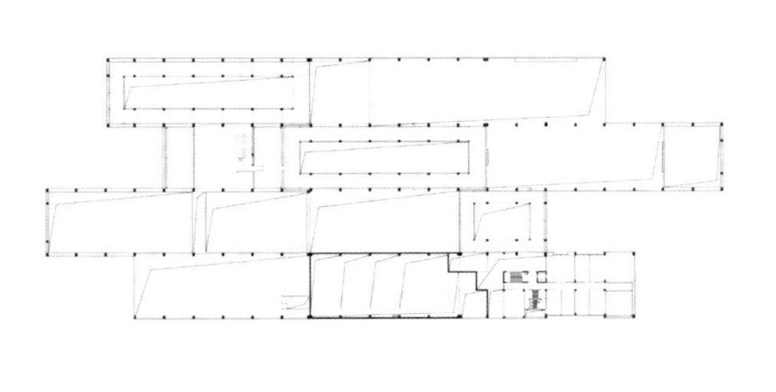

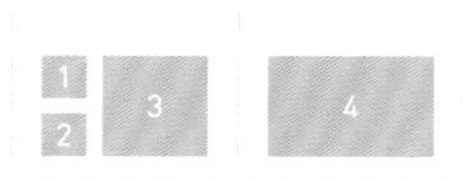

1—3. 良渚博物馆内景
4. 良渚博物馆立面

1—3. Interior Courtyards and Varandans
4. Exterior of the Liangzhu Museum

“现实启发显示。”

关于建筑师

有人说奇普菲尔德是“反扎哈•哈迪德”的建筑师，因为他反对“偶像建筑学”，自称既有现代主义信念，又有保守主义作风：与其冒着风险“想要使每样事物都能够引人入胜，似乎能够改变世界一样”，不如尊重事物本身的含义和以往经验的印记。简言之，他反对浮夸的形式主义，因此其建筑风格朴素低调，张弛有度，体现了英国人的实用和经验主义。然而英国的建筑业也被金钱和潮流所笼罩，尽管奇普菲尔德曾以伦敦为基地，但在自己祖国的竞标中却连连失利，虽起步于伦敦的三宅一生商店等作品，但直到近年才获得肯特郡的特纳当代美术馆等大项目。相比之下，他的名作大多数都在海外，比如西班牙的美洲杯帆船赛馆、德国的现代文学博物馆和柏林新博物馆等。良渚博物馆是奇普菲尔德在中国最早的作品之一，如今他的工作室已从伦敦、柏林、米兰开到了上海。不过，生性谨慎的奇普菲尔德仍会做“噩梦”，那就是“发现自己没有做到最好”。

ABOUT THE ARCHITECT

Some people say that Chipperfield is the "anti-Zaha Hadid" with regards to architecture. This is because he opposes "Idol Architecture", claiming that he represents both modernist beliefs and a conservative style: instead of risking "utilizing everything in order to attract and satisfy an entire audience and seemingly being able to change the world", it is better to respect the meaning of things and the imprint of past experience. In short, he opposes exaggerated formalism. As such, his architectural style is simple and low-key, relaxed and moderate, reflecting British pragmatism and empiricism. However, the British construction industry is also overshadowed by the dual roles of money and trends. Although Chipperfield used to be based in London, he consistently failed to secure biddings in his motherland. Although he first established himself with London's Miyake Life Store in addition to other works, he did not gain the opportunity to work on larger projects such as the Turner Contemporary Art Museum in Kent County until more recent years. By contrast, most of his masterpieces are featured overseas, such as the America's Cup Building in Spain, the Museum of Modern Literature in Germany and the Neues Museum in Berlin. The Liangzhu Culture Museum is one of Chipperfield's earliest works in China. Now his studios have been opened in locations including London, Berlin, Milan and Shanghai. However, Chipperfield's cautious nature results in still constantly facing a "nightmare", namely, "to find that he failed to produce his best work".

大卫•奇普菲尔德
David Chipperfield

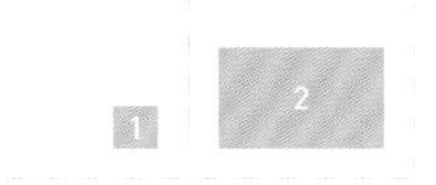

1. 西班牙美洲杯大楼
2. 博物馆中庭水池

1. Veles e Vents
2. Courtyard Water Surface

15
Cultural Complex in Longgang District

深圳龙岗三馆
城市拱门，开向每朵星辰

建筑设计：荷兰麦肯诺建筑师事务所
项目地点：深圳市龙岗区龙城广场
设计时间：2011—2014 年
完成时间：2017 年
建筑总面积：95,000 ㎡
功能：公共艺术馆，科技馆，青少年宫，龙岗书城
商业，地下停车场，公共广场

Arch. design: Mecanoo Architecten
Location: Longcheng Plaza, Longgang District, Shenzhen, Guangdong
Design process: 2011-2014
Completion time: 2017
Total area: 95,000m²
Program: Public Art Museum, Science Centre, Youth Center, Book City
Commercial, Underground Parking, Public Plaza

在深圳龙岗的龙城广场和龙岗商务区之间，一个新的城市文化综合体正在形成。这个由四座建筑构成的综合体包括巨大的龙岗书城、公共艺术中心、青少年宫以及互动科技馆，实现了把龙城广场转变为具有文化氛围的城市中心地带的都市愿景。

狭长的建筑体没有成为公共空间的阻碍，而是像一个连接器一般将周围的区域联系起来。四个建筑体之间围合出不同的开敞空间，形成巨大的、形如拱门一般的通路。其圆润的造型便于人流通过，使西侧的步道和东侧的新商业区形成联系。这些拱形造型、高度和材质均和统一，在不增加多余前后立面建筑的前提下形成了鲜明的视觉参考点和吸引要素，在为以后的扩建留有空间的同时，这些开敞的流线型围合空间自然通风良好，为亚热带城市带来有阴凉的公共领域。

文化综合体的结构立面均由彩色铝板组成，整体建筑看上去充满现代气息。这个倾斜的结构立面包裹着主结构的梁柱系统，让内部空间最大化，形成了无须柱网支撑的“超级中心”。综合体的入口位于巨型“拱门”下方，在这里可以进行各种公共互动，同时也与外界随时保持开放与联系。地面层部分区域向城市开放，两座桥将地面上的建筑与住宅联系起来，呈现出文化与商业和谐共存的城市景象。

A new cultural centre lines the western edge of Longcheng Plaza in the Longgang district, one of the largest suburbs of Shenzhen. Designed to revitalize this park-like square, the complex comprises a vast bookshop, museum for art, youth centre and science centre spread over four volumes.

The elongated building is an urban connector, and triggers – instead of hinders – a connection between the surrounding areas. Large arched passageways between adjacent volumes correspond with pathways on the western side and with infrastructure from a new business district on the eastern side. The round forms of the copper facade support the natural flow of passers-by through the area. By sharing the same form, height and material, the volumes unite as a visually cohesive whole without an apparent front or back facade. The fluid forms of the four volumes channel air currents in a natural way. Visitors and passers-by can shelter from the rain or sun of the local subtropical climate underneath the arches.

The structural facade is cladded in colorized aluminum, which gives the complex a contemporary feel. It also acts as a structural envelope for the beams and the columns of the major structure, in order to create supercores inside with large open spaces.The entrances to the cultural centers are located at these covered squares underneath the contemporary arches, which allow the various interior programmes to extend outdoors. The public appeal generated by this project makes Longcheng Plaza part of a new, vibrant city centre with the cultural complex as a landmark with modern arches and acts as a "Supercore".

1. 一层平面图

1. First Floor Plan

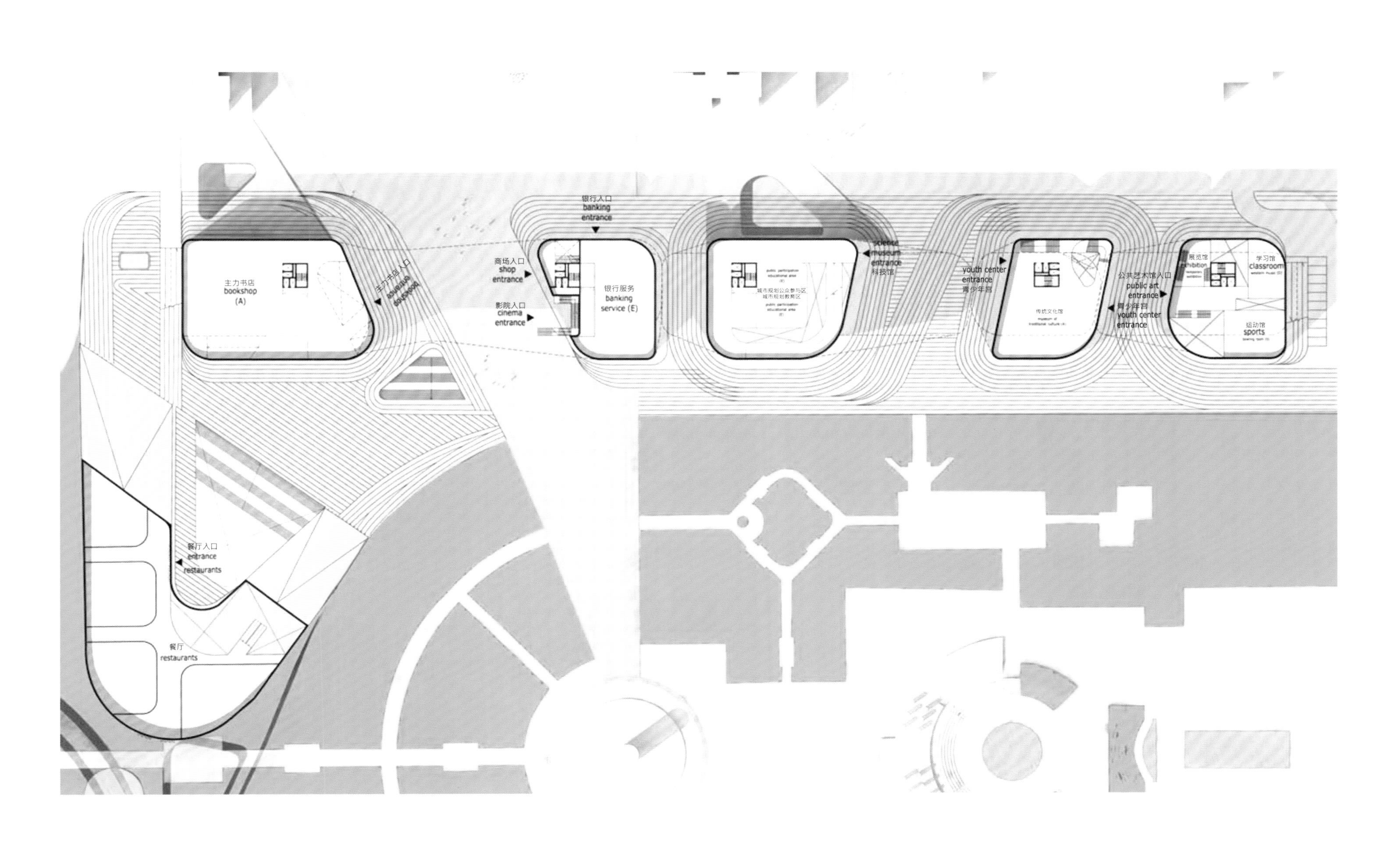

“建筑不只是智性或观念的活动，也不是视觉的表现，而是美学形式与情感的交流。”

关于建筑师

荷本到深圳接手龙岗三馆项目时是带着女儿从荷兰坐火车经西伯利亚来的，可见她的个性像她的设计一样线条鲜明硬朗但不从形态上张扬。她 1984 年大学毕业时在赢得一项集合住宅竞赛后旋即与同伴创立的麦肯诺（名称源于一种金属组合玩具，寓意为“快乐地建造”），如今已是荷兰五大建筑师团体之一。早年凭借代尔夫特理工大学图书馆跻身主流设计圈的她，在 2007 年高雄卫武营艺术文化中心的竞赛中获胜，成为与扎哈 • 哈迪德比肩的明星，并正式进入亚洲，于 2011 年因其规划“充分体现了将龙城广场转变为具有文化氛围的城市中心地带的都市愿景”而被深圳市龙岗区政府选中。麦肯诺提倡“有机设计”，以自然生态和可持续性为基础，将人文、科技、趣味等因素融入每个方案，举重若轻地诠释光和美，让建筑通过对形态和情感的组织来触动所有的感知。

ABOUT THE ARCHITECT

When Mrs. Houben took over the Cultural Complex project in Longgang District, she took a train all the way from Siberia with her daughter. Just like her lifestyle, the designs of Houben's are vivid with personalities yet not aggressively showing off in appearance. Mecanoo (a metal toy set meaning "happy building"), founded by Houben and her colleges in 1984 right after her graduation and successful competition of a mass housing project, has now become one of the 5 major Dutch architecture associations. Her early project, library of Technische Universiteit Delft, had brought her into the main stream design realm; and in 2007 she was awarded for her Wei-Wu-Ying Center for the Arts in Kaohsiung, Taiwan. Houben then became the star architect as hot as Zaha Hadid and officially recognized in Asia, hence she was selected by Longgang Government to conduct a major planning in regards to revitalize the Longcheng Plaza into a CBD with cultural vibe. Mecanoo encourages "organic design" that bases on sustainable development with a focus on natural habitat, integrating humanities, science and other fun elements into each project. The Mecanoo team had brought great attentions to light and aesthetics, communicating their perceptions through objective form and abstract emotions behind architecture.

法兰馨 • 荷本
Francine Houben

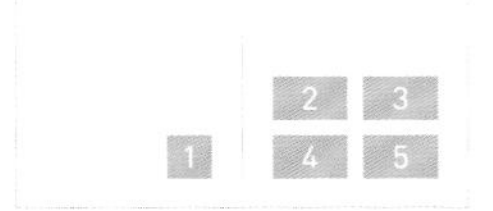

1. 荷兰代尔夫特理工大学图书馆
2. 立面形态处理
3. 丰富的“公共穿越空间”
4. 穿越“城市拱门”
5. 仰望“城市拱门”

1. Library Delft University of Technology
2. Elevation Design
3. Dynamic "Urban Connectors"
4. Through the "Modern Arch"
5. Looking up the "Modern Arch"

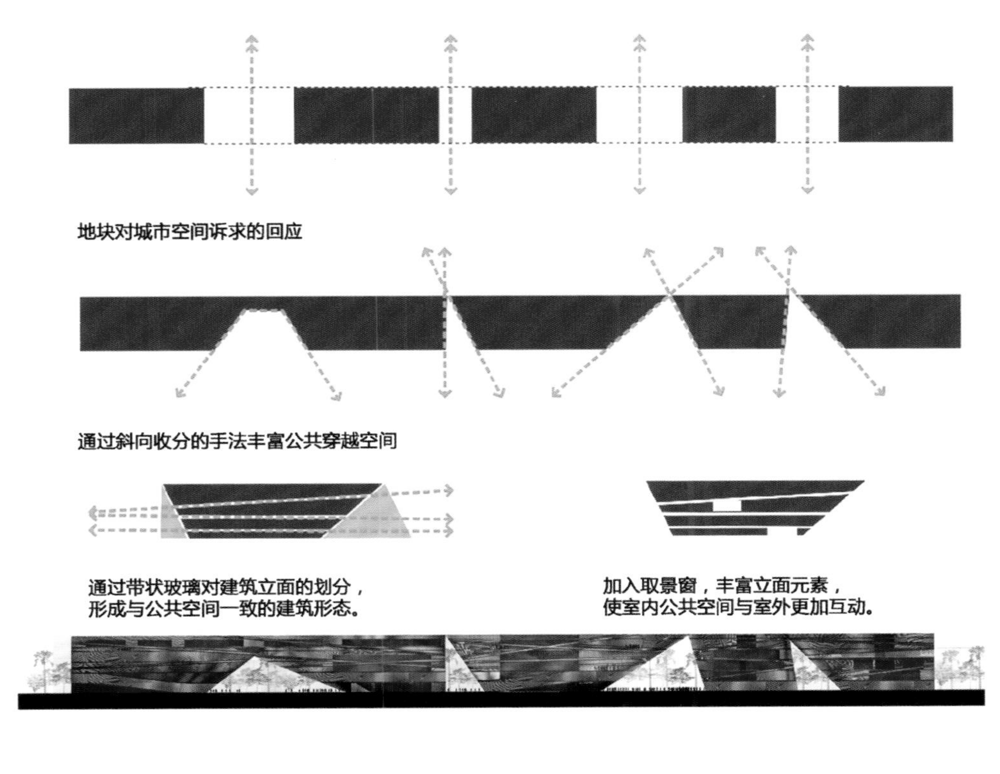

立面形式的处理手法上，采取活跃、彰显的设计态度，
通过建筑空间与材料自身品质来打造深圳龙岗区的形象气质。

1890

16

Zhang Zhidong Museum

张之洞博物馆

方舟、天空、大地、过去、未来的动态解读

建筑设计：里伯斯金工作室

项目地点：武汉

完成时间：2018 年

建筑面积：7,240 ㎡

功能：博物馆

Archi. design: Studio Libeskind

Location: Wuhan

Completion time: 2018

Total area: 7,240m²

Program: Museum

张之洞博物馆在武汉的旧钢铁厂基地上建成，它以一栋建筑景观来纪念张之洞这位对武汉现代化工业进程发展具有重要意义的历史人物，并着眼于本地工业以及武汉市的过去、现在和未来。这些故事环环相扣，成为武汉关键历史的一部分。在建筑物内有许多对光线的设计理解和展示，凸显出天空与大地，人与意义。利用充满多样性和灵动性的解读，帮助参观者理解这些抽象性的概念。

“方舟”的建筑形式受武汉传统中国建筑古代象征主义影响，可以被看作在黄鹤楼上所使用过的屋顶形式“飞檐”的放大。博物馆的外形概念既是一种空间意符，也为跨越武昌、汉口和汉阳提供了水上通道。建筑物本身被提升至空中，它将成为旁边新区域未来发展的门户。也正因此，它为各种活动提供了公共空间，这也符合这座建筑物的雕塑理念。使用全钢结构的初衷是忠于武汉的钢铁遗迹，因为这座建筑是在汉阳铁厂原址上新建的，这座建筑的整体结构采用钢铁建筑技艺以凸显其背后雄伟的工业力量。

这座博物馆是通向未来公园及工业遗址区的入口。高耸的建筑形式具有让人惊叹的轻巧钢制外壁，似乎在向积极的未来延伸。两个展厅空间将建筑物高高举起，占据突出位置的一个展厅用于公众活动，另一个是员工办公室和场馆工作区。框架式支撑的结构由巨大的钢柱支撑，用来表现曾经在此遗址上出现过汉阳铁厂的规模和坚固性。建筑物及花园将采用可持续技术完成，也将成为未来可持续发展的典范。

Located at the site of Wuhan's old steelworks in Wuhan, the Museum of Zhang ZhiDong is dedicated to this prominent and historical local figure who helped modernize Wuhan through the Industrial Revolution. The museum is designed to recall the city's industrial past, while looking to the future. The manipulation and understanding of light inside the museum are dynamic in regard to the meaning between sky, earth and human-beings, providing a guideline of how these abstract relationships could be revealed in a diversified scheme.

The form of the architecture is described as an "Arc", was influenced by the symbolism of ancient Chines architecture form in Wuhan, which could be understood as an enlargement of the upturned eaves of Tower of Yellow Crane. At the apex of the building, a lattice opening the structure allows for views out towards the city of Wuhan. It is symbolic in space arrangement and also a pathway between Wuchang, Hankou and Hanyang. The elevated building would become a gateway for future development of the new district adjacent to the site, and provide public space for various activities as a sculptural backdrop. The steel-structure is a reinforcement in remembrance of the old industrial heritage scene of Wuhan, as the building is erected on top of the old Hanyang Ironwork Factory.

As an entry of the park and the future industrial heritage zone, the extraordinary light-weighted steel facade extends into the sky. The two main exhibition halls are elevated, one used as public activities and the other is used as offices and workzone. Mega-steel-structure, acts as a framework of the supporting system, is also a symbol of the past, representing the once-glorious vast ironwork factory and solidity of the industrial scene. The garden and the building are all built under sustainable design scheme for a future with environmental concern.

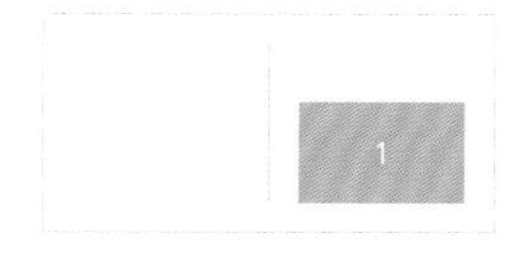

1. 里伯斯金张之洞博物馆手绘，2012

1. Zhang Zhidong and Modern Industrial Museum-Sketch by Libeskind, 2012

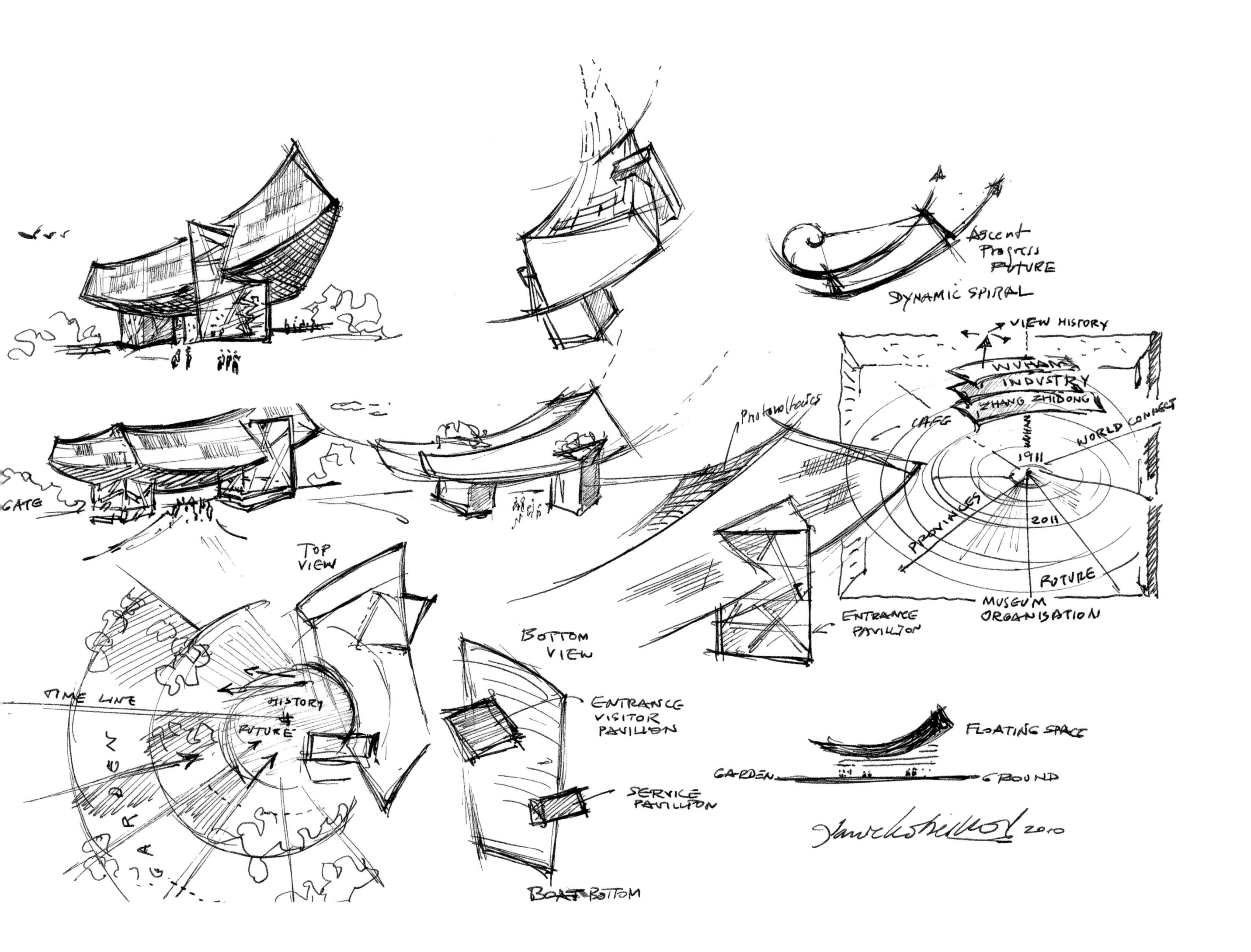

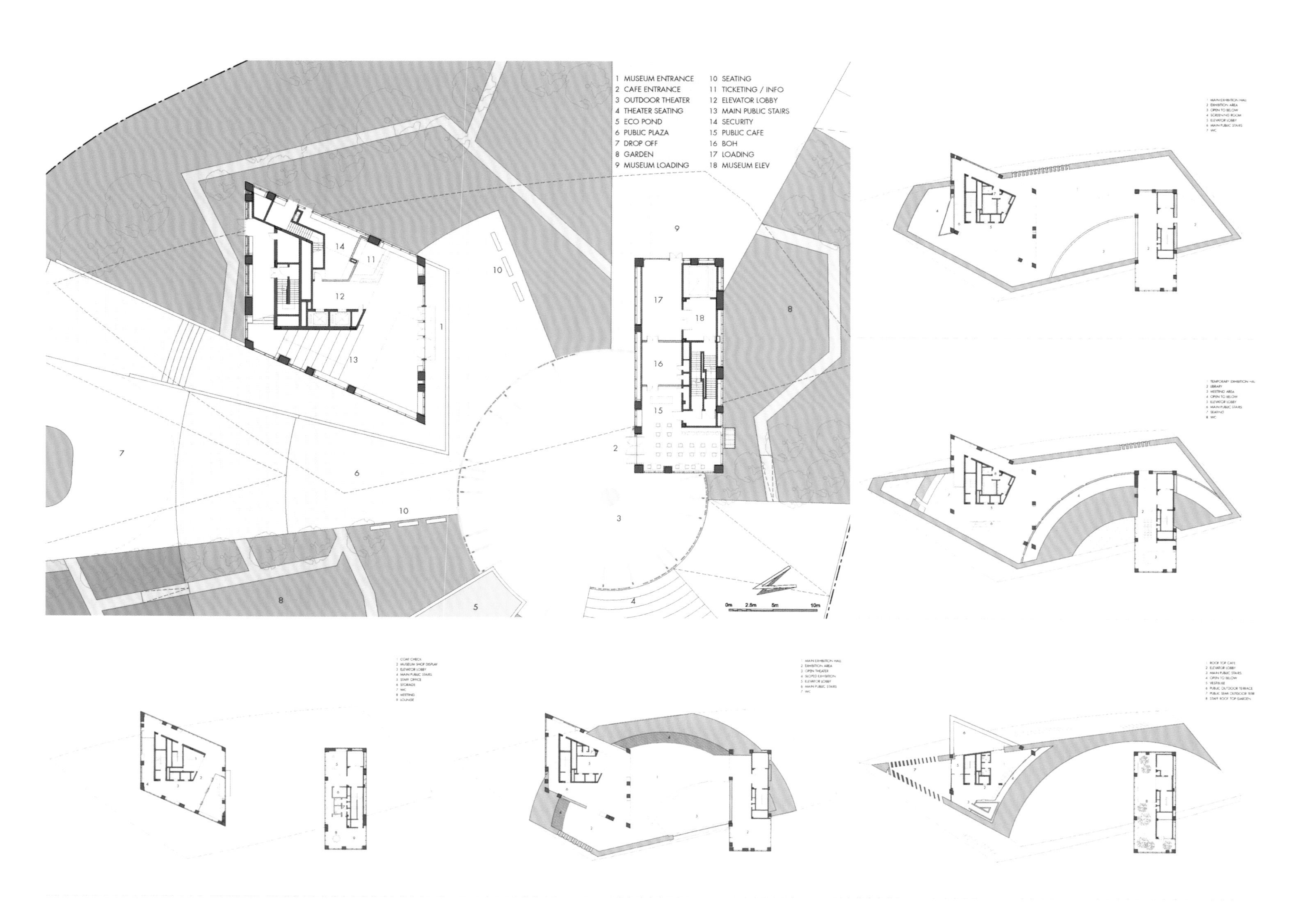
1 MUSEUM ENTRANCE
2 CAFE ENTRANCE
3 OUTDOOR THEATER
4 THEATER SEATING
5 ECO POND
6 PUBLIC PLAZA
7 DROP OFF
8 GARDEN
9 MUSEUM LOADING
10 SEATING
11 TICKETING / INFO
12 ELEVATOR LOBBY
13 MAIN PUBLIC STAIRS
14 SECURITY
15 PUBLIC CAFE
16 BOH
17 LOADING
18 MUSEUM ELEV
0m 2.5m 5m 10m

1. 总平面图
2—6. 各层平面图
7—10. 立面图

1. Masterplan
2—6. Floor plans
7—10. Elevations

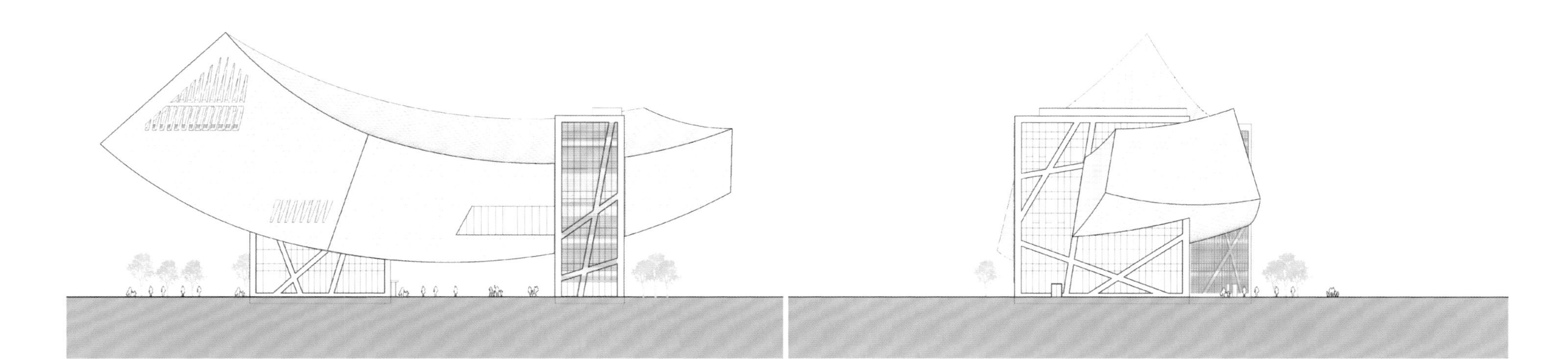

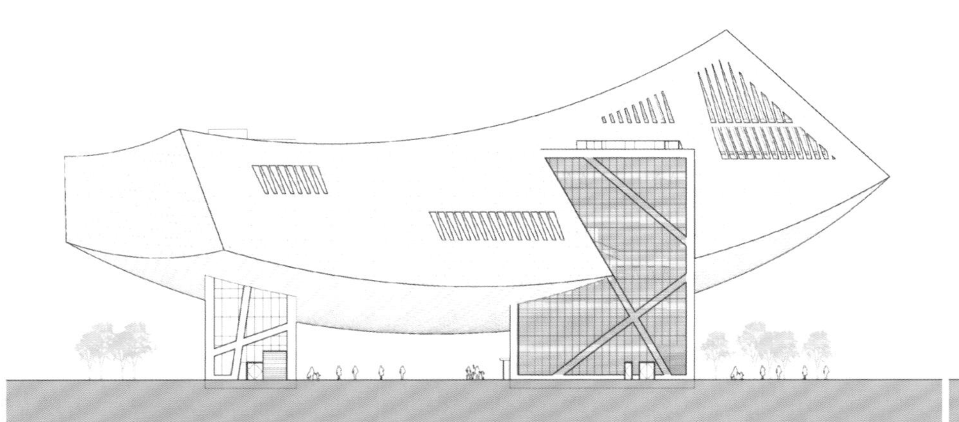

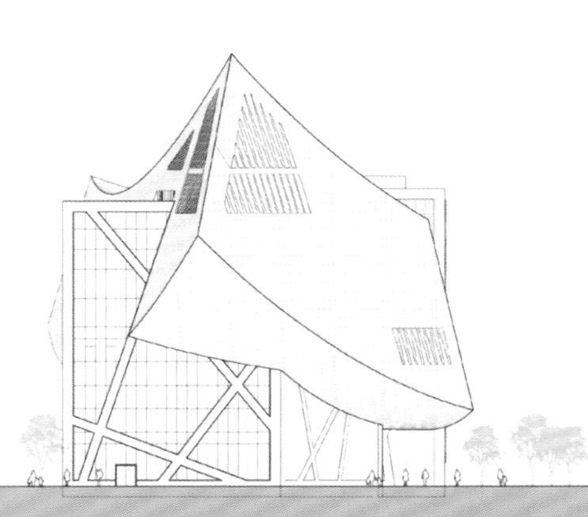

“如果你不相信有更美好的未来，你就不能从事建筑设计工作。”

关于建筑师

伟大的建筑“能说出灵魂深处的精彩故事”，这是美籍波兰犹太人里伯斯金的名言。他父母是大屠杀的幸存者，1957 年举家迁往以色列，两年后定居纽约。里伯斯金从库珀联盟学院和埃塞克斯大学毕业后长期是默默无闻的教书匠，到八十年代后期才在米兰通过参加竞赛开启他的执业生涯。1998 年，52 岁的里伯斯金建成第一件作品费利克斯 • 努斯鲍姆博物馆，而次年竣工的柏林犹太博物馆则让他蜚声世界。他相继完成曼彻斯特帝国战争博物馆、丹佛美术馆、皇家安大略博物馆等一系列文化及商业建筑的设计或扩建，在“9•11”事件后获选为纽约世贸中心重建项目的总规划师，被誉为“博物馆之王”。他受万科邀请来华设计的张之洞与近代工业博物馆位于汉阳铁厂旧址，用全钢结构和“方舟”形式来向武钢和武汉的过去、现在和未来致敬。“记忆与纪念是建筑的根本”，里伯斯金相信，“只有直面伤痛和历史，才能最终治愈它，遗忘从来都不是解决之道”。

ABOUT THE ARCHITECT

Great architecture "can tell wonderful stories that reach into the depths of the soul". This is the famous saying of American Polish Jew Libeskind. His parents were survivors of the genocide, and in 1957 his family moved to Israel, after which they subsequently settled in New York two years later. After graduating from Cooper Union and Essex University, Libeskind was a relatively unknown pedagogue for long period of time. It was not until the late 1980's that he embarked on his career in Milan by participating in competitions. In 1998, the 52-year-old Libeskind completed his first work, the Felix Nussbaum Museum. The Jewish Museum in Berlin, which was completed the following year, earned him prestige and fame throughout the world. He successively completed the design or expansion of a series of cultural and commercial buildings, including the Manchester Imperial War Museum, the Denver Museum of Art and the Royal Ontario Museum. After "9·11",he was selected as the chief planner of the World Trade Center reconstruction project in New York and was known as the "King of Museums". He was invited on behalf of Vanke to undertake the design of the Zhang Zhidong and Modern Industrial Museum, which is located at the old site of the Hanyang Ironworks. The project pays tribute to the past, present and future of Wuhan and WISCO by applying an all-steel structure and "ark" form. "Memory and commemoration are the foundation of architecture," Libeskind believes. "Only by facing pain and history directly, can we ultimately find healing. Forgetting is never the solution."

丹尼尔 • 里伯斯金
Daniel Liberskind

1. 柏林犹太人博物馆
2. 张之洞博物馆

1. Jüdisches Museum Berlin
2. Zhang Zhidong and Modern Industrial Museum

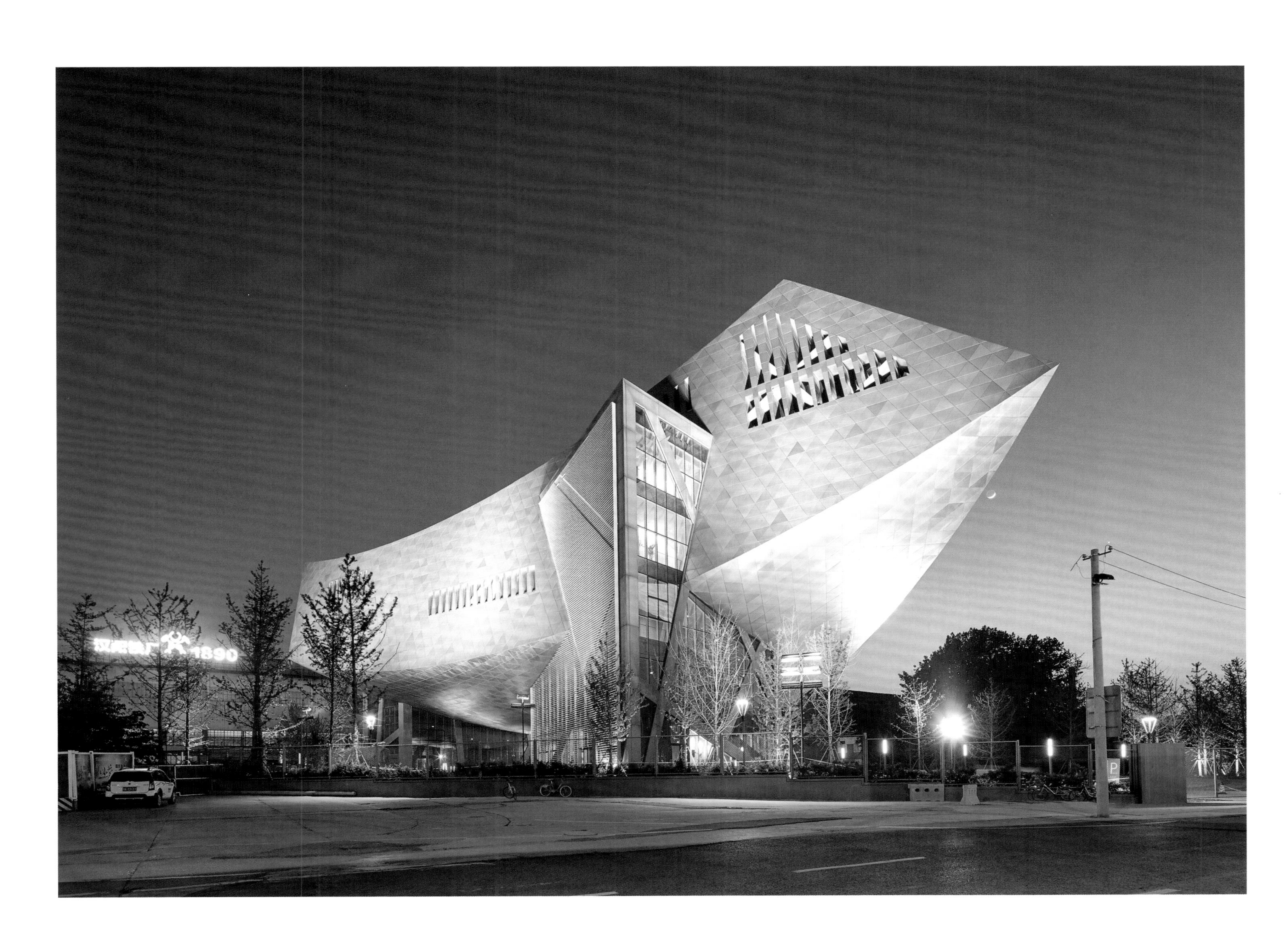

汉阳铁厂 1890
禁止车辆通行

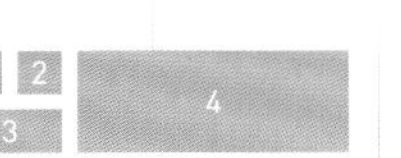

1—4. 张之洞博物馆实景图

1—4. Zhang Zhidong and Modern Industrial Museum at Different Viewpoints

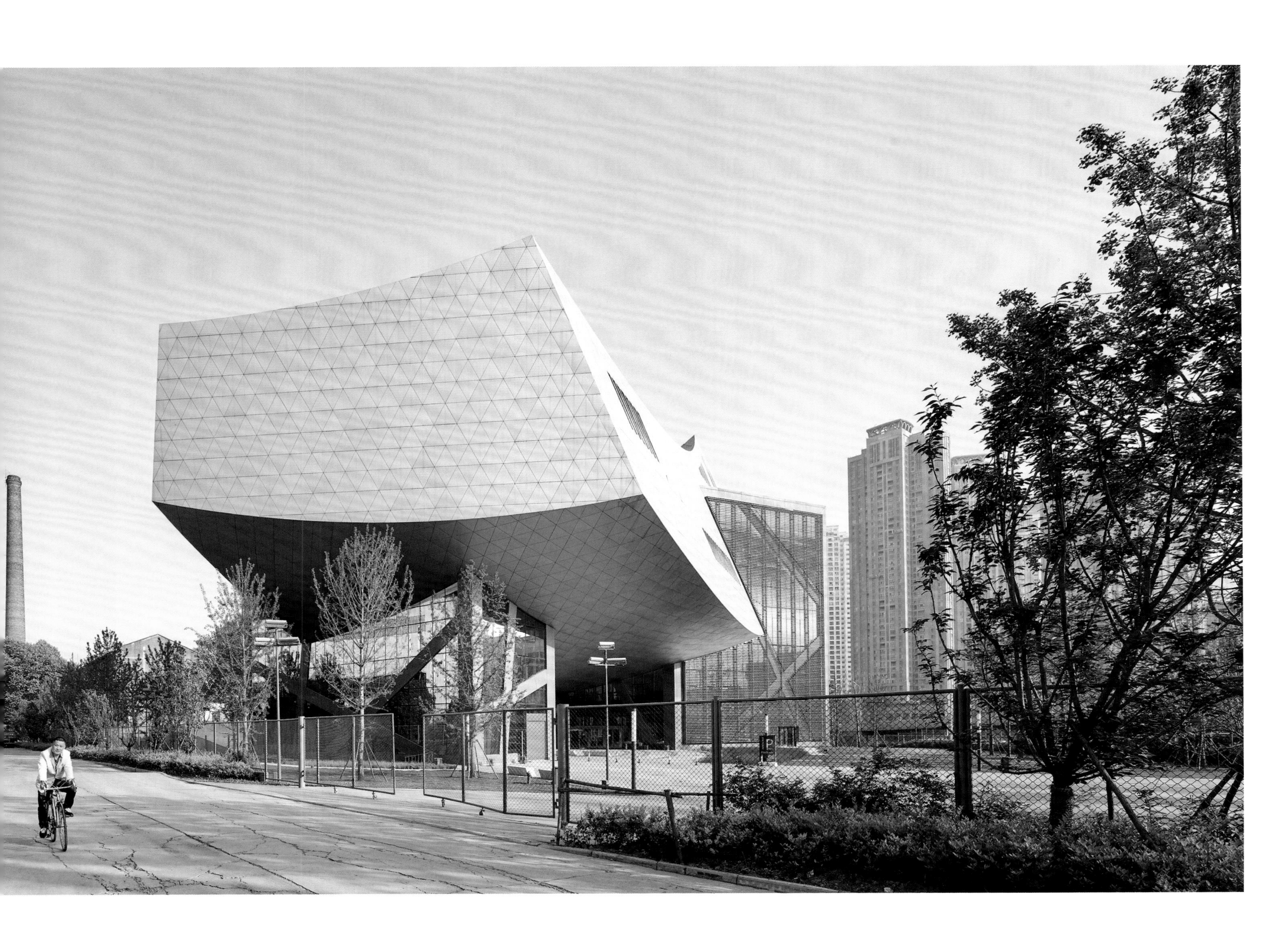

PART E Urban Agriculture
都市农业

17 Beijing No.4 High School (Fangshan Campus)

北京四中房山校区
自然与开放的学校

建筑设计：OPEN 建筑事务所
项目地点：北京市房山区长政南街 6 号
完成时间：2014 年
建筑总面积：57,773 ㎡
功能：学校

Arch. design: OPEN Architecture
Location: No. 6 Changzheng South Street, Fangshan District, Beijing
Completion time: 2014
Total area: 57,773m²
Program: School

北京四中房山校区

这个占地 4.5 公顷的新建公立中学位于北京西南五环外的一个新城的中心，是著名的北京四中的分校区。新学校是这个避免早期单一功能的郊区开发模式、更加健康和可持续的新城计划重要的一部分，对新近城市化的周边地区的发展起着至关重要的作用。

创造更多充满自然的开放空间的设计出发点——这是今天中国城市学生所迫切需要的东西，加上场地的空间限制，激发了我们在垂直方向上创建多层地面的设计策略。学校的功能空间被组织成上下两部分，并在其间插入了花园。上部建筑和下部空间垂直并置，并在“中间地带”（架空的夹层）以不同方式相互接触、支撑或连接，这既是营造空间的策略，也象征了这个新学校中正式与非正式教学空间的关系。

下部空间包含一些大体量、非重复性的校园公共功能，如食堂、礼堂、体育馆和游泳池等。每个不同的空间，以其不同的高度需求，从下面推动地面隆起成不同形态的山丘并触碰到上部建筑的“肚皮”，它们的屋顶以景观园林的形式成为新的起伏开放的“地面”。上部建筑是根茎状的板楼，包含了那些更具重复性的和更严格的功能，如教室、实验室、学生宿舍和行政楼等。它们形成了一座巨构，有扩展、弯曲和分支，但全部连接在一起。在这个巨大的结构中，主要交通流线被拓展为创建社交空间的室内场所，就像一条河流，其中还包含自由形态的“岛屿”，为小型的群组活动提供半私密的围合空间。教学楼的屋顶被设计成一个有机农场，为 36 个班的学生提供 36 块实验田，不仅让师生有机会学习耕种，还对这片土地曾作为农田的过去留存敬意。

两种类型的教育空间之间的张力，及其各自包含的丰富的功能，造就了令人惊讶的空间的复杂性。为每类不同的功能所做的适合其个性的空间，使得这个功能繁杂的校园建筑具备了城市性的体验。 与一个典型的校园通常具有的分等级的空间组织和用轴线来约束大致对称的运动所不同，这个新学校的空间形式是自由的、多中心的，可以根据使用者的需求从任意可能的序列中进入。空间的自由通透鼓励积极的探索并期待不同个体从使用上的再创造。希望学校的物理环境能启发并影响当前中国教育中一些急需的变化。

这个项目是中国第一个获得绿色建筑三星级认证的中学［其标准超过 LEED 金级认证］。为了最大化地利用自然通风和自然光线，并减少冬天及夏天的冷热负荷，被动式节能策略几乎运用在设计的方方面面中，大到建筑的布局和几何形态，小到窗户的细部设计。地面透水砖的铺装和屋顶绿化有助于减少地表径流，两个大型雨水回收池从操场的地下收集宝贵的雨水灌溉农田和花园。地源热泵技术为大型公共空间提供了可持续能源，同时独立控制的 VRV 机组服务于所有单独的教学空间，确保使用的灵活性。整个项目使用了简单、自然和耐用的材料，如竹木胶合板、水刷石（一项正在消失的工艺）、石材和暴露混凝土等。

在中国当前的环境下，可以说最迫切的问题和挑战就是人与社会之间以及人与自然之间的关系，而教育承担着巨大的责任。面对这些问题，我们以这个新校区项目作为试金石和力所能及的回应。

1

1. 屋顶平面鸟瞰图

1. Rooftop Birdeye View

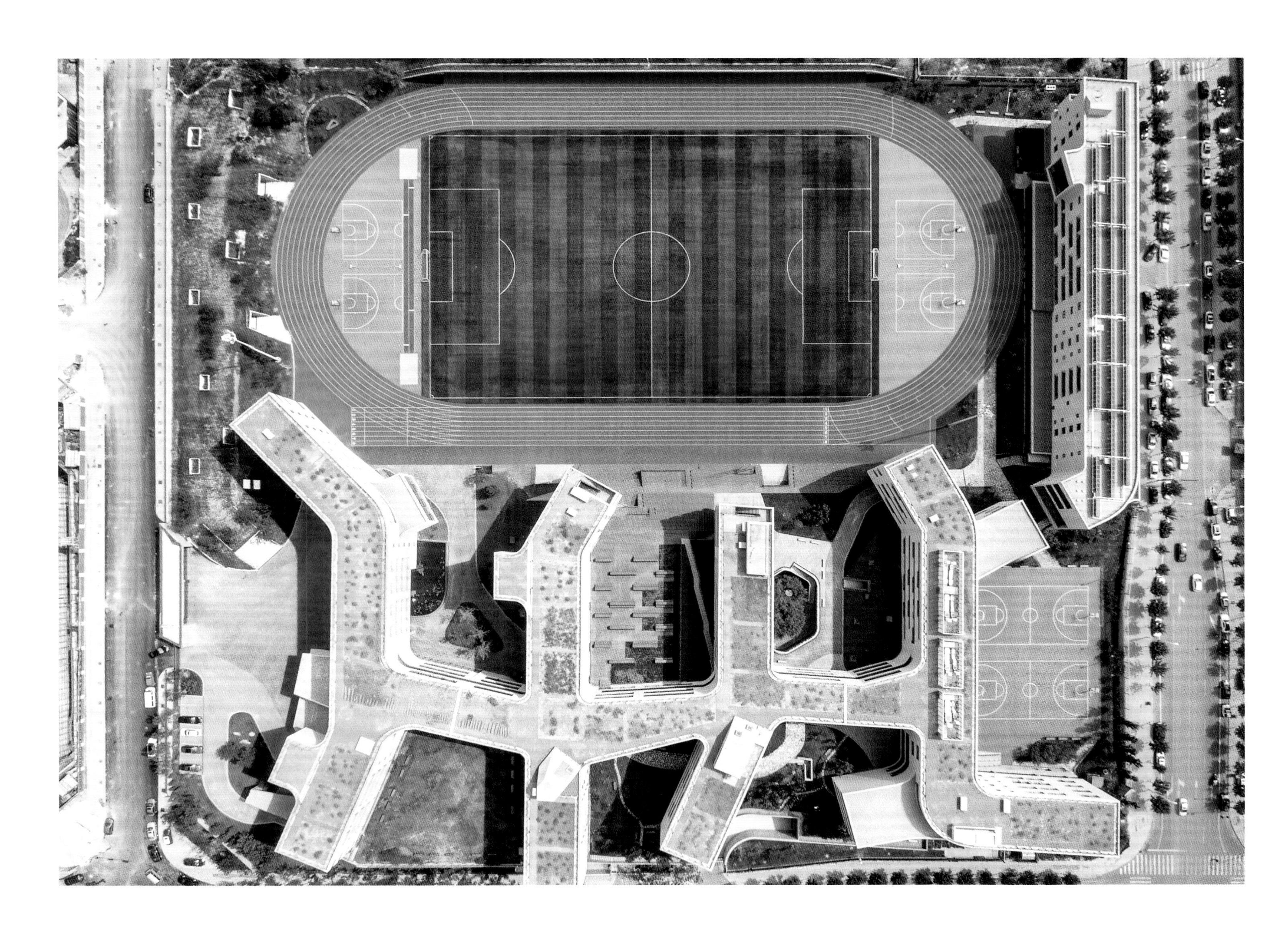

Situated in the center of a new town just outside Beijing's southwest fifth ring road, this new public school on 4.5 hectares of land is designed as the branch campus for the renowned Beijing No.4 High School. As an important piece in a grand scheme to build a healthier and self-sustainable new town, avoiding problems of the earlier mono-functional suburban developments, the school is vital to the newly urbanized surrounding area.

The intention of creating more open spaces filled with nature, something that urban Chinese students today desperately need, combined with the space limitations of the site, inspired a strategy on the vertical dimension to create multiple grounds, by separating the programs into above and below, and inserting gardens in-between. The juxtaposition of the resultant upper and lower building, connected at the "middle-ground"in various ways, is as much an interesting spatial strategy as a signifier of the relationship between formal and informal educational spaces in the new school.

The lower building contains large and non-repetitive public functions of the school, such as the canteen, the auditorium, the gymnasium, and the swimming pool. Each of these spaces with their varying height requirements, push the ground up from below into various mound shapes that touch the belly of the upper building; their roofs in the form of landscaped gardens become the undulating new open "ground". The upper building is a thin rhizome shaped slab that contains the more repetitive and rigid programs of classrooms, labs, dormitories and administration. Its mega form extends, bends, and branches, but all connected together. The main circulation spine within this mega structure is widened to allow rapid foot traffic during class breaks. It also accommodates some semi-enclosed spaces for small group activities, like a river with organic shaped islands. The roof-top of the upper building is designed to be an organic farm, with 36 plots for the 36-classes of students in the school, providing students the chance to learn the techniques of farming, and also paying tribute to the site's pastoral past.

The contrast between the two types of educational spaces and the rich mix of programs within create a surprising spatial complexity. With the unique character for each different space, an urban experience is created within this complex of education facilities. Unlike a typical campus with hierarchical spatial organization and often clear axis to organize more or less symmetrical movements, this new school is free form and meant to have multiple centers that can be accessed in any possible sequences. It is a place with a free spirit that encourages explorations and awaits reinventions by different individuals. Hopefully the physical environment can inspire and initiate some much needed changes in the education system of China today.

This project aims to be the first triple-green-star rated school in the country (a standard that exceeds LEED Gold). In order to maximize natural ventilation and natural light, and minimize heat gain during summer and heat loss in the winter, passive solar strategies are adopted in almost all aspects of the design, from the planning of the building geometry all the way to the details of the window design. Permeable ground surface paving and expansive green roofs helps to minimize surface run-off, and three large underground water retention basins collect precious rain water from the athletics field for irrigation of the farms and gardens. A geothermal ground-source heat pump provides a sustainable source of energy for the large public spaces, whilst independently controlled VRV units serve all the individual teaching spaces to ensure flexible operation. Throughout the project, simple, natural, and durable materials such as bamboo plywood, pebble dashing (a vanishing technique), stone, and exposed concrete are used.

In the contemporary Chinese context, arguably the most pertinent issue and challenge is that of the relationship among individual, society, and nature. Education bears great responsibilities. It is to these issues that this new campus project aspires to be both a touchstone and a response.

1. 剖面图
2. 立面图

1. Section
2. Elevation

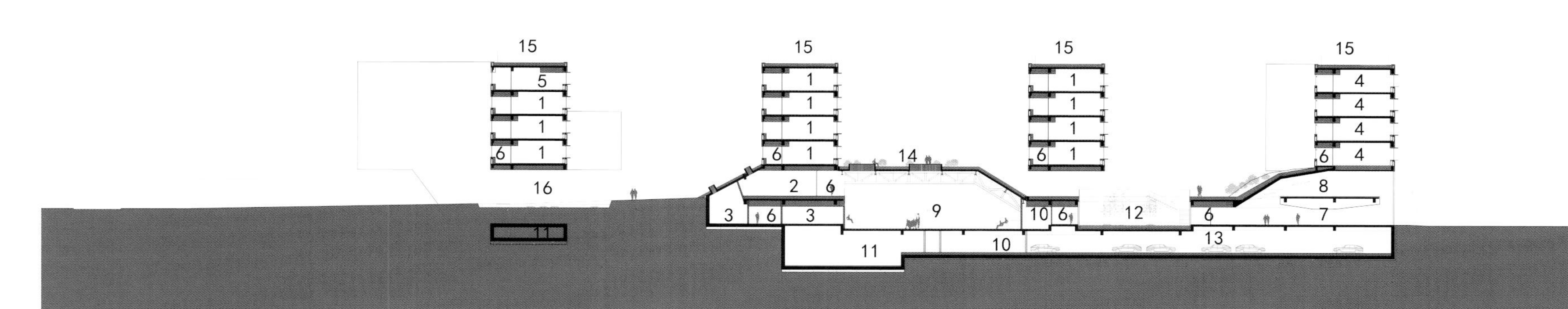

1. 教室
2. 音乐教室
3. 技术教室
4. 实验室
5. 图书馆
6. 走廊
7. 学生餐厅
8. 教师餐厅
9. 风雨操场
10. 储存室
11. 设备用房
12. 竹园
13. 车库
14. 庭院
15. 农田
16. 水池

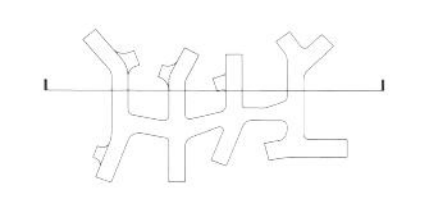

剖面图 0 2 5 10 20M

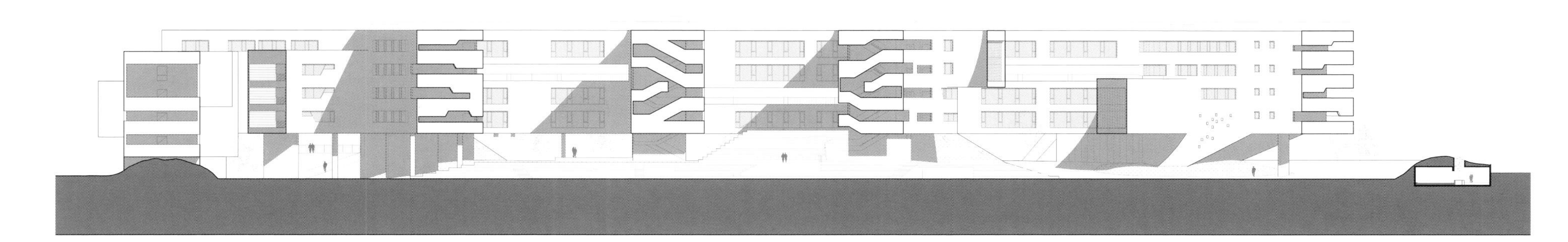

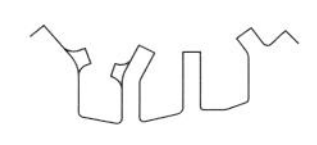

东立面图 0 2 5 10 20M

“在当下的时代和社会背景下，建筑以其创新的力量，去影响并改变人们和他们的生活方式，同时在建造与自然之间达成平衡。”

关于建筑师

“我最关心的是如何在私营开发的项目上，提供一些公共空间。我每个项目都是在做这个概念。”2006 年身为霍尔（Steven Holl）事务所合伙人的李虎回到北京时深感于社会共享资源的匮乏，从而成为北京当代 MOMA、深圳万科中心等以开放和环保著称、极具社会意义的建筑背后的灵魂人物。2010 年他与黄文菁创立的 OPEN 建筑事务所在设计北京四中房山校区的竞赛中胜出，通过营造教科书式的田园学校来重塑人与自然、人与社会的关系。注重建筑的精神性和实用性的李虎近年来接的主要是艺术和教育这两类他眼中社会最需要的项目，但其实他对城市人性化的思考早在他在美国莱斯大学读硕士时用聚焦北京北二环问题的小电影做毕业设计就开始了，而他的建筑中常见的底层架空等主题则源于他还在清华读大三时接受的柯布西耶的影响。对较真率直、热爱生活的李虎来说，做建筑和做人是一回事，“都反映了你的世界观和价值观”。不体现建筑师个人表达或信仰的建筑是空洞的，然而李虎所追求的不仅仅是人性解放，也是美好和谐的共同生活。

ABOUT THE ARCHITECT

“My major concern is to find a way to provide public space in private development projects. All of my projects are experimenting this idea.” Li Hu, formal partner of Steven Holl Architects, had felt the lack of public social resource sharing in Beijing when he came back to China in 2006. He then became THE man behind the scene of the Beijing MOMA and Shenzhen Vanke Center, both of which have great social significance for their openness and concern in environment. In 2010, Li Hu, together with his co-founder, Huang Wenjing had won the competition under the name their studio “OPEN” in designing the Fangshan Campus for Beijing No. 4 High School. The project seeks a text-book style of picturesque campus to re-establish the relationship between human, nature and society. Li Hu concerns in both spirit and practicality in architecture, he has been emphasizing on art and education projects, which he understands as the most-needed categories in the current society. When he was studying Master in Rice University, Li Hu had chosen a self-produced documentary on problems of North Second-Ring Road in Beijing as his graduation thesis, reflecting his thoughts in humanity in cosmopolitan. In Li Hu's project one can usually see a raised floorplan design, which could be traced back to his early collage years in Tsinghua University when Li was greatly influenced by Corbusier. He is a straight-forward guy with passion in life, and a little stubborn sometime. “Your conception of world and value would both show in the way you live and the way you design (architecture).” Li Hu would say, and he believes that architecture should reflect designer's personal belief, in his case, a harmonic life with liberation of humanities.

李虎 & 黄文菁

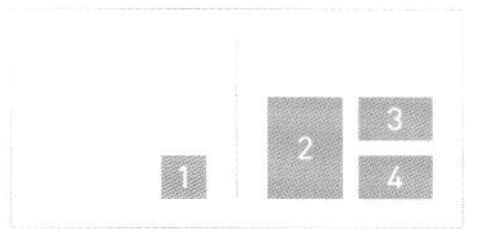

1. 清华大学海洋中心
2—4. 学校实景

1. Tsinghua Ocean Center
2—4. School Photos

PART F Mixed City, 3-Dimensional City
混合城市，立体城市

18 Vanke Yun City Planning

万科云城规划

超大尺度城市多元、集群、分时序发展的案例

总体城市设计和景观设计：都市实践

项目地点：深圳

设计时间：2013 年至今

用地面积：约 38 公顷

建筑总面积：1,330,000 ㎡

功能：城市设计，建筑设计，景观设计

Overall Urban Design and Landscape Desig: URBANUS

Location: Shenzhen

Design process: 2013—present

Site area：Approx. 38 Ha

Total area: 1,330,000m^2

Program: Urban Design, Building and Landscape Design

在留仙洞万科云城设计公社的项目中，都市实践整合了中西方城市发展的经验，为深圳超大尺度城市发展设立了多元、集群、分时序发展的案例。从整体开发的角度来看，在研究世界各地的总部基地园区演变过程中，都在不断寻找城市尺度空间与土地价值最大化之间的平衡，而最终提出单一开发商综合体的模式。在万科留仙洞项目中，都市实践希望避免中国现有总部园区没有空间整体性及功能整体性的问题，并利用万科的软实力及整体开发难得的机会，成就下一个中国经典案例。在整体城市设计的前提下，提出集群设计的落实方法，目的是在可控的城市框架中，因为各个设计师的独立思考，出现多元性。设计统筹成为城市设计落实及深化的关键。

In the Liuxiandong (LXD) Vanke Yun City project, URBANUS had integrated development and design models from Chinese and western cities, and finally proposed the large scale block complex model for homogeneous development in Shenzhen. In LXD project, URBANUS advocates a precinct integrated development model that overtakes the other cases. In the developing process of headquarter parks all over the world, people have been seeking for the best balance between urban space and land value, this model will be the key to realizing various new technologies and dreamed city lives. The design must avoid issues that occur to China's existing headquarter parks, such as lacking spatial and programmatic integrity. The project should make use of Vanke's soft power and seize the rare opportunity of carrying out integrated development. A collaboration with the newly minted UDL (Urban Design & Landscape team), the Vanke LXD project represented a new innovation in approach, whereby URBANUS would not only craft the base control plan and the initial urban design vision, but would assist in the "curation" of the individual architectural designs within the framework of controllable urban structure, which then becomes a key role in realizing and intensifying in further urban planning.

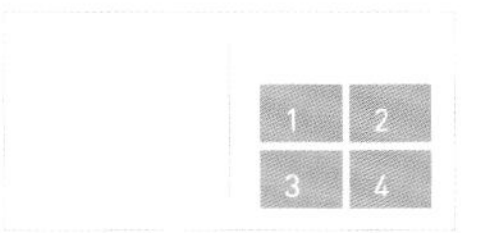

1. 孟岩手　　留仙洞城市设计
2. 多样，混合功能
3. 留仙洞总平面图
4. 建筑空间类型

1. Sketch by Meng Yan
2. Diversified Programs
3. Masterplan
4. Architectural Space Type

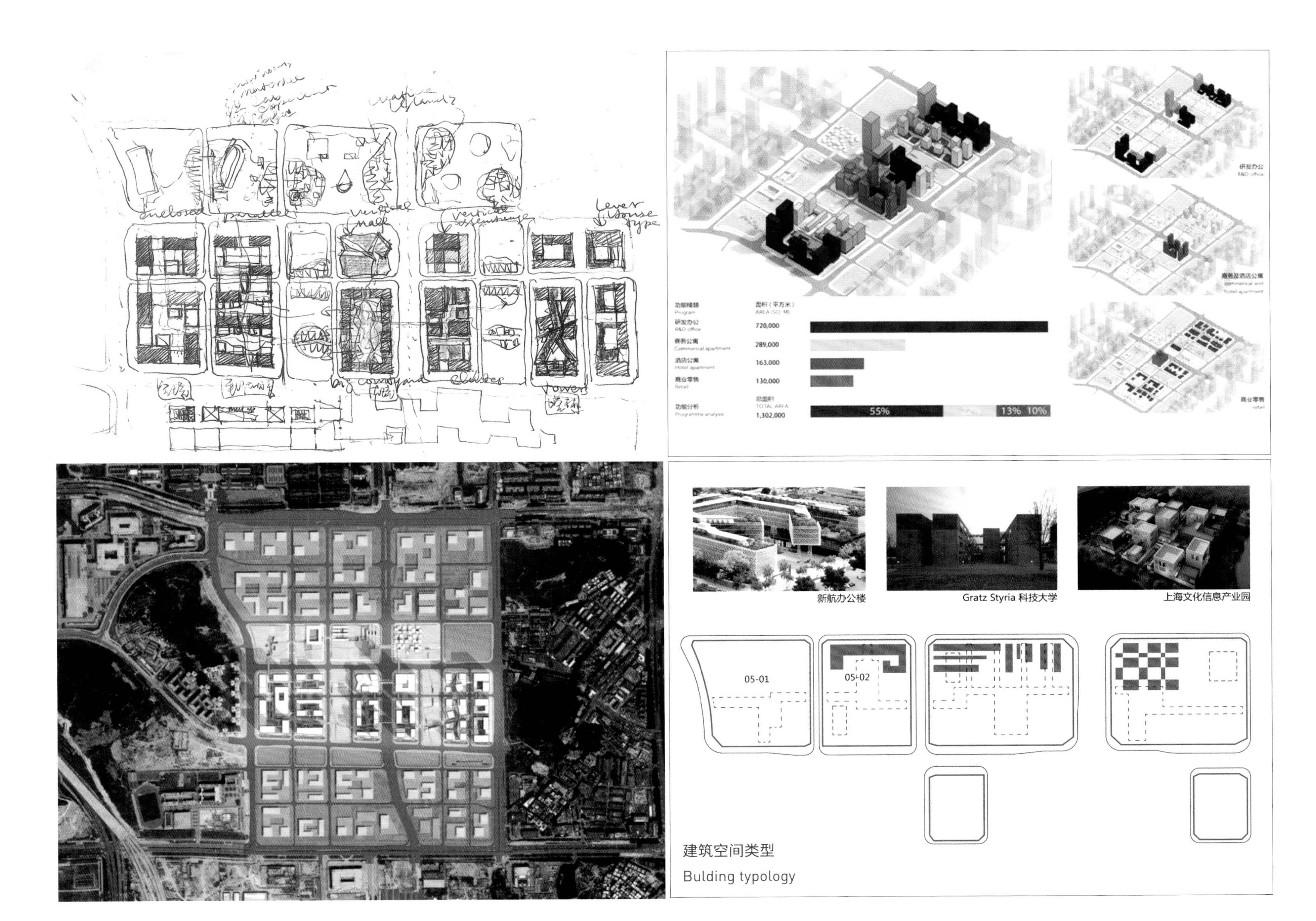

“我们始终关注现实，要为新世纪不断产生的城市问题寻找新的解决方案。我们相信只有如此，才能真正做到具有全球化视野的同时，又深深植根于本土的当代建筑实践。”

关于建筑师

都市实践（URBANUS）是当今中国最具影响力的建筑师团队之一，由刘晓都、孟岩和王辉创建于 1999 年。目前在深圳、北京设立有分支机构。它既是一个机构，更是一种理念，旨在从广阔的城市视角和特定的城市体验中解读建筑的内涵，紧扣中国的城市现实，以研究不断涌现的当下城市问题为基础，致力于建筑学领域的探索。都市实践的工作以其对城市的社会和历史结构的敏锐，整合复杂的城市环境里潜在的空间和社会文化资源，准确找到有效的应对策略而受到国际关注。

都市实践自成立以来参与了近 400 个重要的建筑和规划设计项目，2010 年成立研究部专注于城市研究。针对中国城市现状，积极推进对创意产业模式、高密度开发、后城中村现象等课题的研究。

ABOUT THE ARCHITECT

Founded in 1999, under the leadership of partners Liu Xiaodu,Meng Yan and Wang Hui, Urbanus is recognized as one of the most influential architecture practices in China. More than a design practice, Urbanus is also a think tank. It aims at formulating architectural strategy from the urban environment in general and the ever changing urban conditions. Urbanus always focuses on urban realities in China and seeks architectural solutions based on its research of the emerging urban problems. Urbanus has completed a series of design projects featuring various scales and functions, the works of Urbanus have drawn international attention due to the company's sensitivity to urban historical and social structure, integration of potential resources of space and society, and effectiveness in responding to the complicated urban environment.

Urbanus has completed over 400 important architecture design and planning projects. The Urbanus Research Bureau (URB), established in 2010, which is primarily concerned with urban research, focuses on the contemporary urban China phenomenon to conduct a series of research projects, including: creative city development, post-urban village development, typologies for hyperdensity and others.

都市实践合伙人：
孟岩、刘晓都、王辉

URBANUS
Meng Yan, Liu Xiaodu, Wang Hui

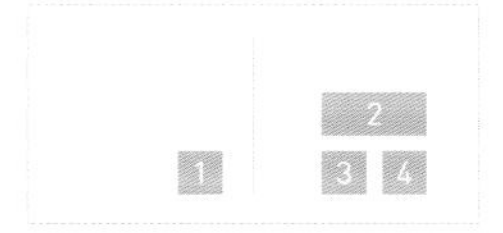

1. 雅昌艺术中心
2—4. 未来街景示意图范例

1. Artron Art Center
2—4. Future Urban Landscape Precedents

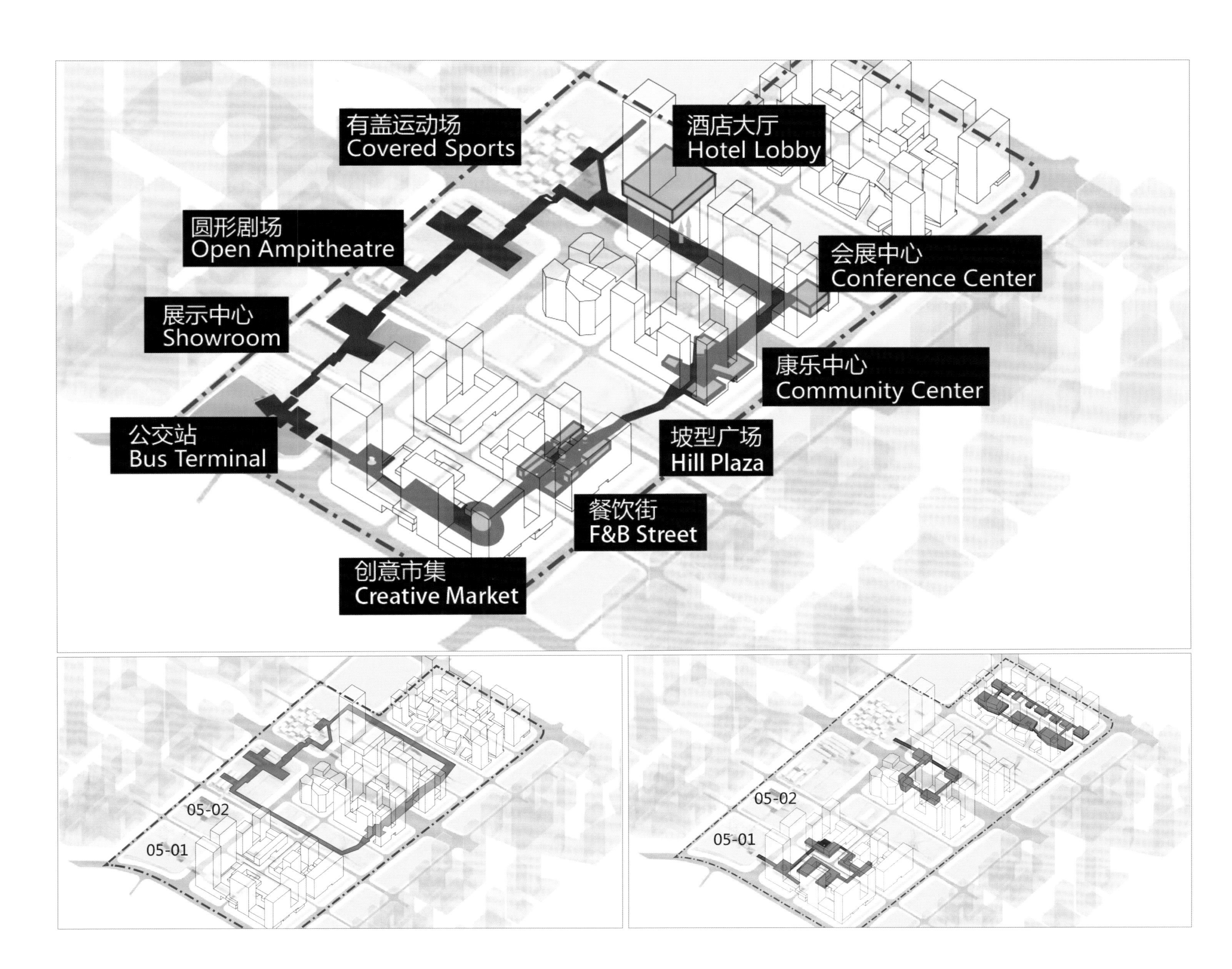

整个城市设计的基本概念是借鉴纽约市的条状结构，连接南侧与北侧的已开发地区及地铁站点。贯穿南北的城市肌理令整个地块分割成八条南北向的街区，其中六块为建设区，而两块为绿廊。根据纵向的街区分区，在城市设计阶段已经为每个分区设立了各种混合功能的配比。

这种城市设计方法，基本上杜绝了单一功能街区所带来的各种问题，并令各个分区在功能上能够有所区分。本项目最为重要的整个绿地空间的体系被作为功能复合型进行设计开发，并在绿地空间中设法加入建设面积，形成商业功能，进一步激活绿地空间的生命力。

北侧的绿地空间是整个项目的亮点，也是整个项目建设上的领航员。作为第一阶段的建设，绿地空间的地下空间承接了大量的城市内容，在功能上作为整个总部园区的孵化区，承载未来创新企业的幼苗，并聚集了大量的城市创新人员，为基地带来前所未有的活力。而当其余阶段的建设在未来若干年建设完毕后，我们更希望这些机构已经具有更强的实力，进驻总部园区，继续充当城市各个阶段发展的催化剂，并传承者留仙洞总部园区的基因。

The city planning of LXD project is referenced to the strip-structure of New York City, connecting the existing established zones and metro stations of the northern and southern sides. The city, penetrating through the north-south axis, is divided into 8 north-south oriented street stripes, within which 6 are future-construction zones and 2 are green-belts. During the early urban planning stage, the mixed functions within each of the vertical street divisions had been pre-designed with a proper ratio.

Such planning method had overcome the possible problems due to monolithic street function, and predicted different usages in zonings. The most important space, the "green belts", are designed into a complex mixed-use urban space, with carefully added architecture design and commercial spaces, the green belts are to become the most energetic zones in the planning.

For example, the green space at the northern side is the key to the project, it also acts as the curator of the entire planning scheme. As a part of the phase I construction, the underground of the green space had accommodated a large number of city functions, functioned an incubator zone for the whole LXD industrial zone for future creative participants to host their new-established companies. The dynamic atmosphere that brings would further develops into opportunities in business power, and inherits the headquarter gene of LXD area that would acts as a catalyst to various phases of future LXD city development.

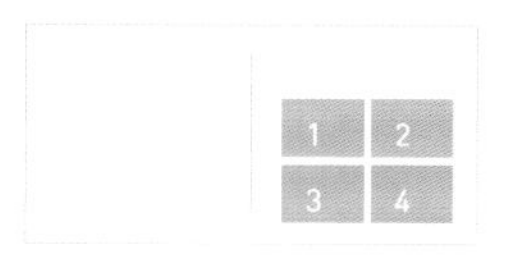

1—4. 效果图

1—4. Renders

城市系统——外部空间联系

City System—External Connection

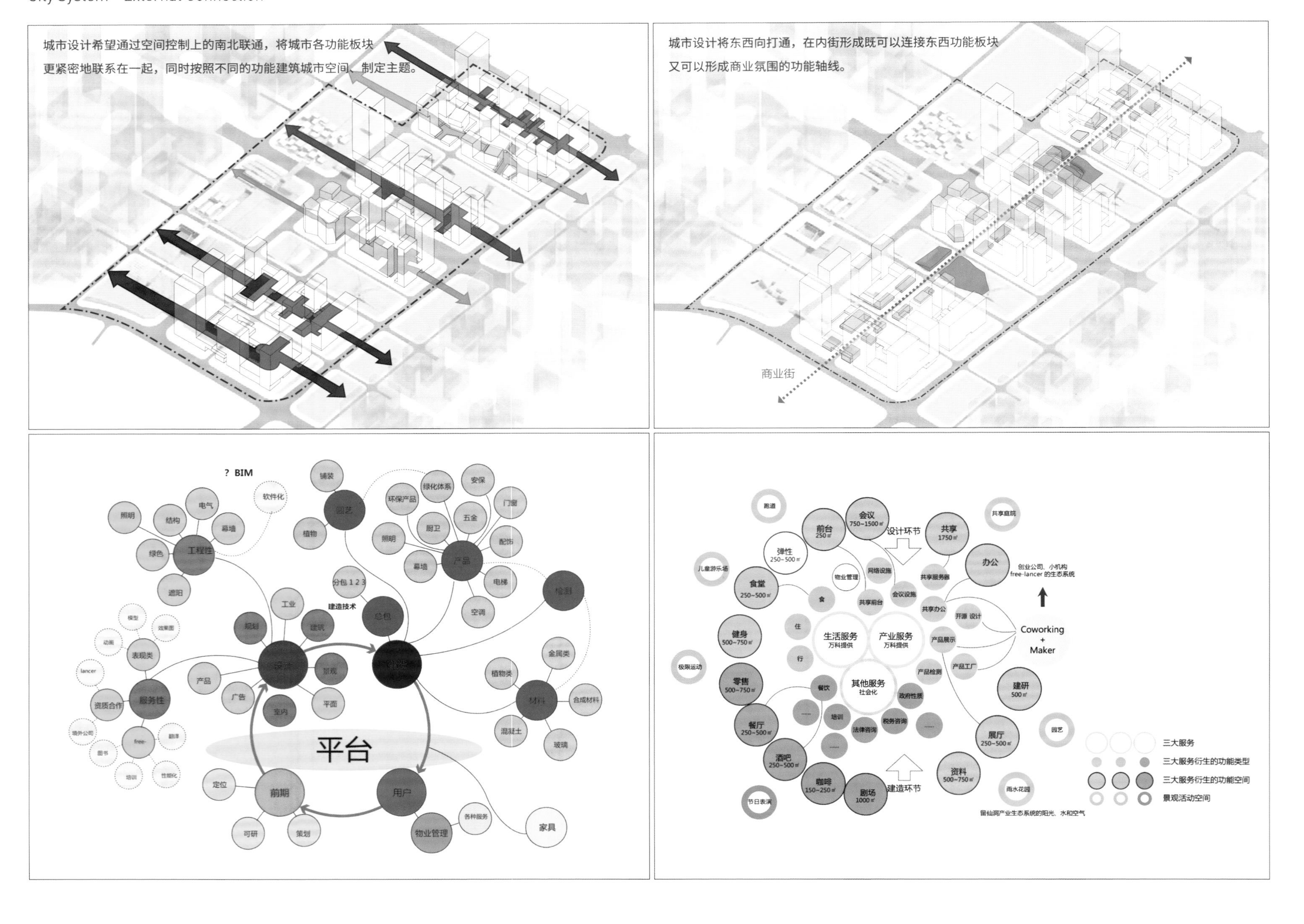

1. 城市南北向轴线
2. 城市东西向轴线
3. 企业模式整合
4. 空间系统原则与功能分布
5. 空间控制原则

1. North—South Axis
2. East—West Axis
3. Integration of Industry Mode
4. Space System Rule and Programe
5. Space Control Rule

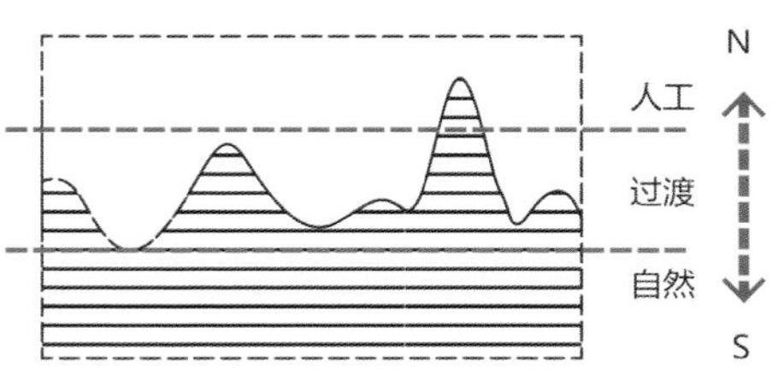

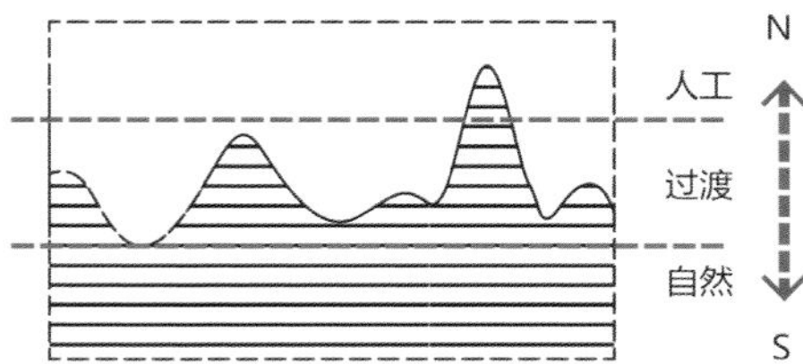

从自然到人工的趋势

1. 整体公园形态从自然过渡到人工，确保所有地块南侧的东西向连续性
2. 屋顶绿化需要在地块中占据三分之二以上面积
3. 过渡区域形态宜以较为分散的形态出现，丰富内部公共空间

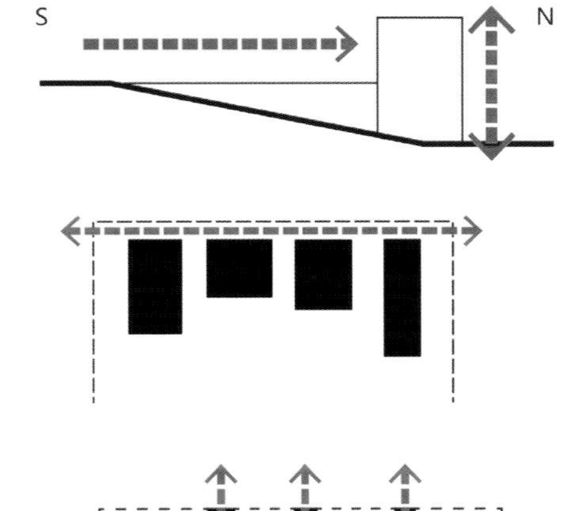

地块建筑界面

1. 建筑屋顶南北向需要满足标高的连续性，标高不宜大于一层高差（5m）
2. 确保地块南侧从道路进入地块的便利性
3. 地块北侧统一建筑红线且东西向排齐
4. 建筑退线3米形成与负一层相同标高道路，贯通东西向，各建筑需要在流线上与该标高相连接

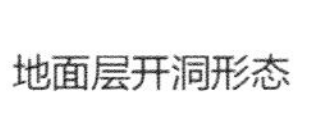

地面层开洞形态

1. 遵循南侧东西向连接形态，中部明显东西向负一公共空间东西向形态
2. 北侧南北向与外部连通
3. 围绕中部公共廊道空间的形态可丰富多变

微地形趋势

1. 整体公园形态从自然过渡到人工，确保所有地块南侧的东西向连续性
2. 确保05-01公交站屋顶道路进入场站屋顶的可能性
3. 确保路径向05-02地上建筑聚拢的可能性，由此形成坡地标高

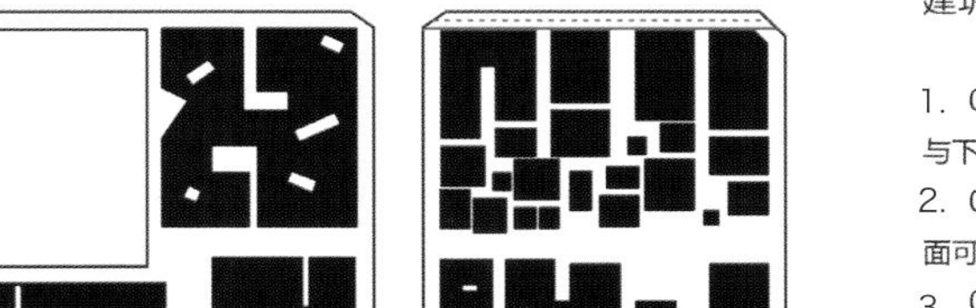

建筑与公共空间图底关系

1. 05-01屋顶平面具有地形起伏，负一层以采光井与下沉庭院为主，屋顶平面具有标高的连续性
2. 05-02南侧屋顶标高具有联系性，同时东西向屋面可以连通
3. 05-02北侧分两部分，靠近公共走廊部分以较为细碎的体量创造空间丰富性，其屋顶可以有一定的连通和高差变化
4. 05-02北侧四栋较为独立的建筑可以有一定的东西向连续性

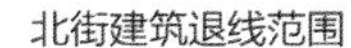

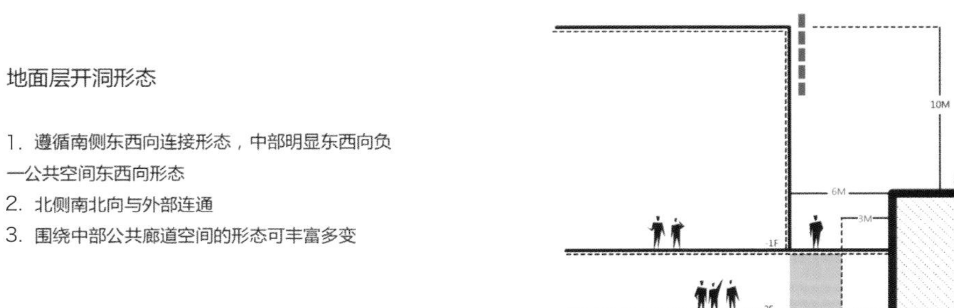

北街建筑退线范围

1. 退线负一层为6米，负二层为3米，负二层墙线与负一层墙线之间区域可考虑挖洞与否
2. 05-02北侧建筑限高10米，需要注意在每栋建筑顶层立面投影线须与建筑用地红线距离为6米，整体四栋建筑顶层立面守齐

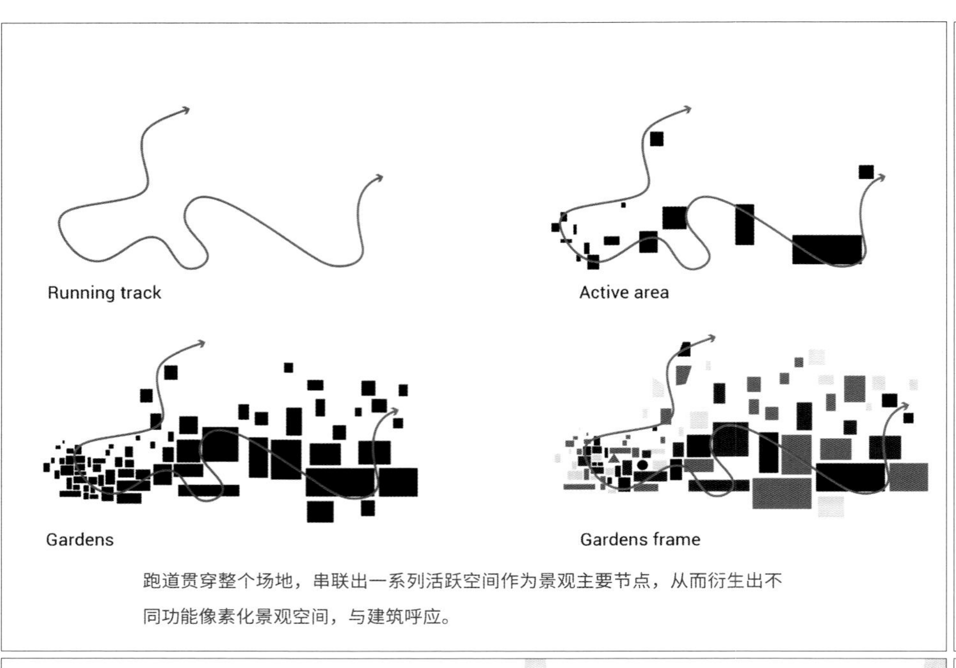

跑道贯穿整个场地，串联出一系列活跃空间作为景观主要节点，从而衍生出不同功能像素化景观空间，与建筑呼应。

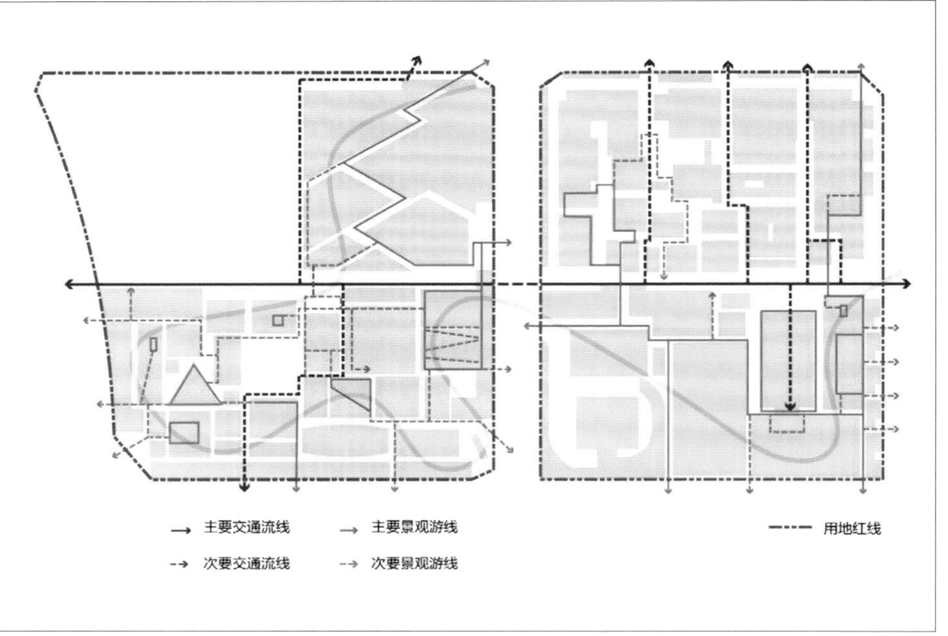

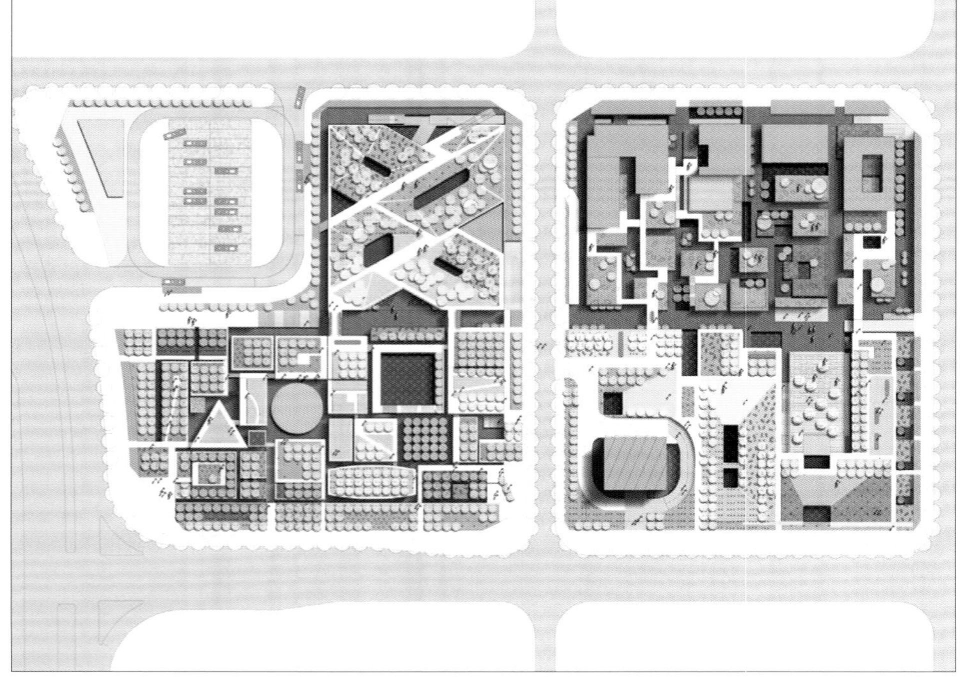

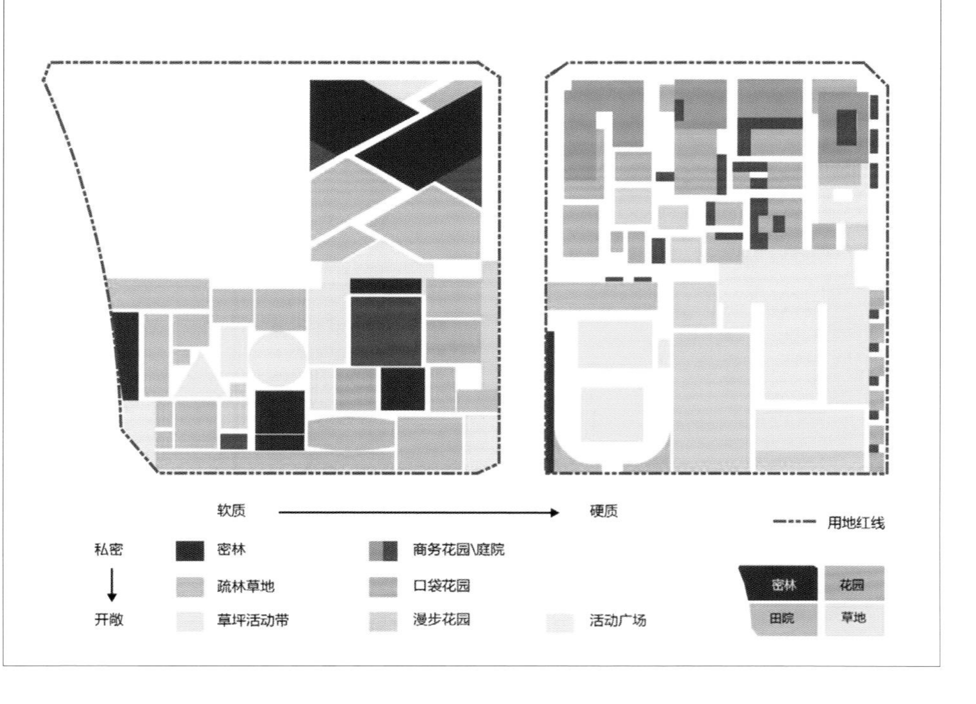

1. 景观概念设计
2. 景观流线分析
3. 景观总平面图
4. 景观功能分析
5. 建筑概念设计
6. 建筑平面图一层
7. 建筑平面图负一层

1. Lanscape Design Concept
2. Main Part Entrance
3. Lanscape Masterplan
4. Lanscape Function Analysis
5. Architectural Design Concept
6. Ground Floor Plan
7. Basement 1 Floor Plan

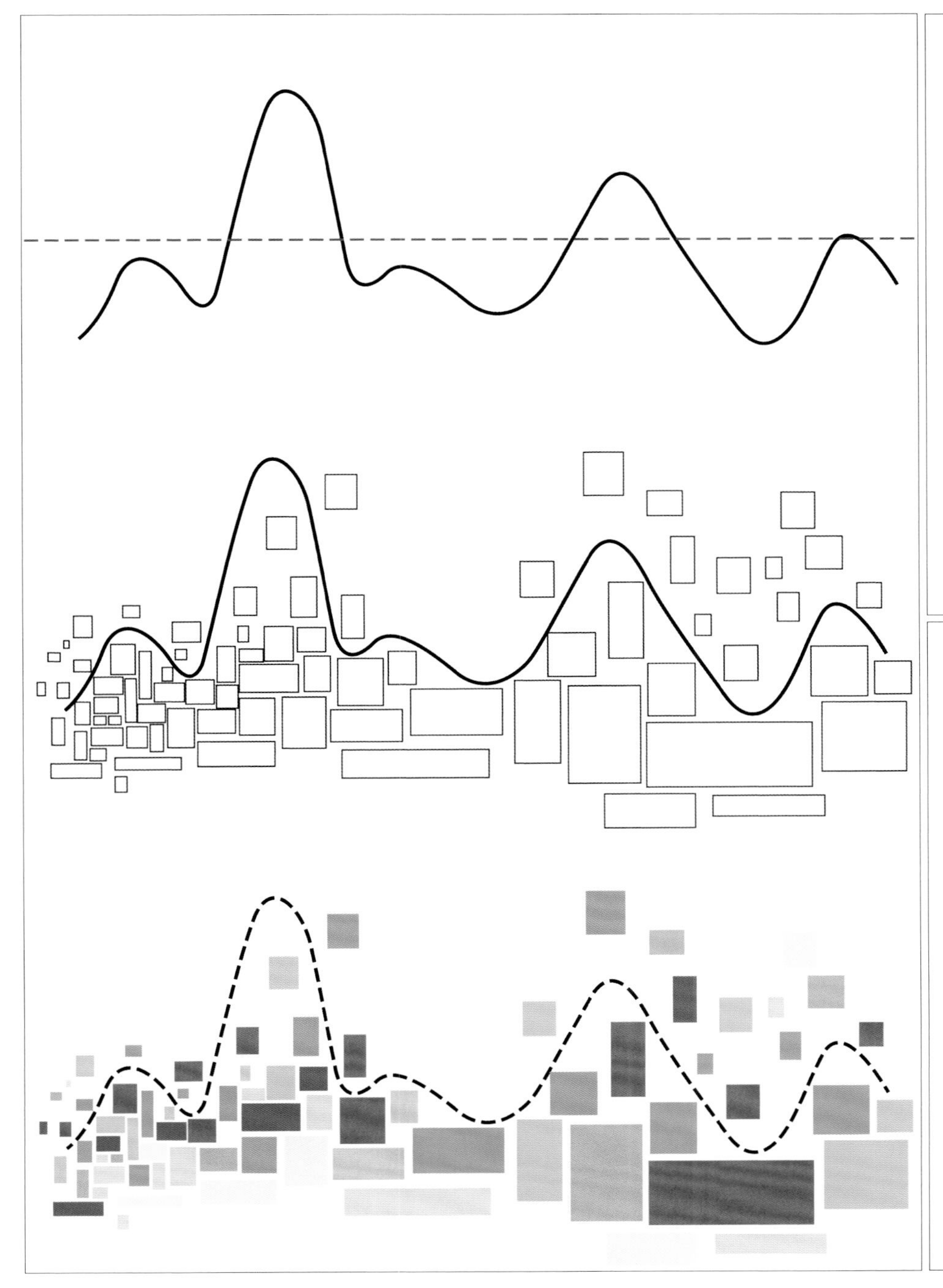

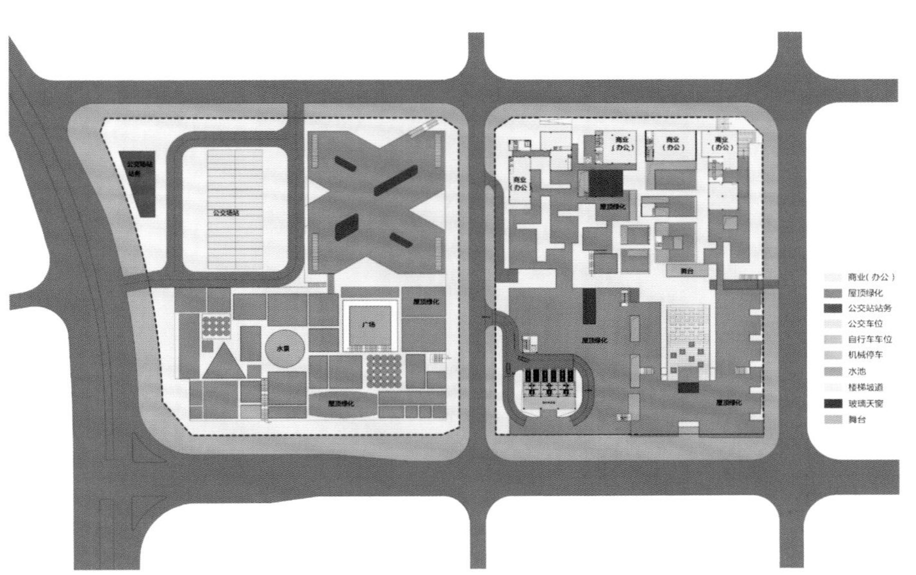

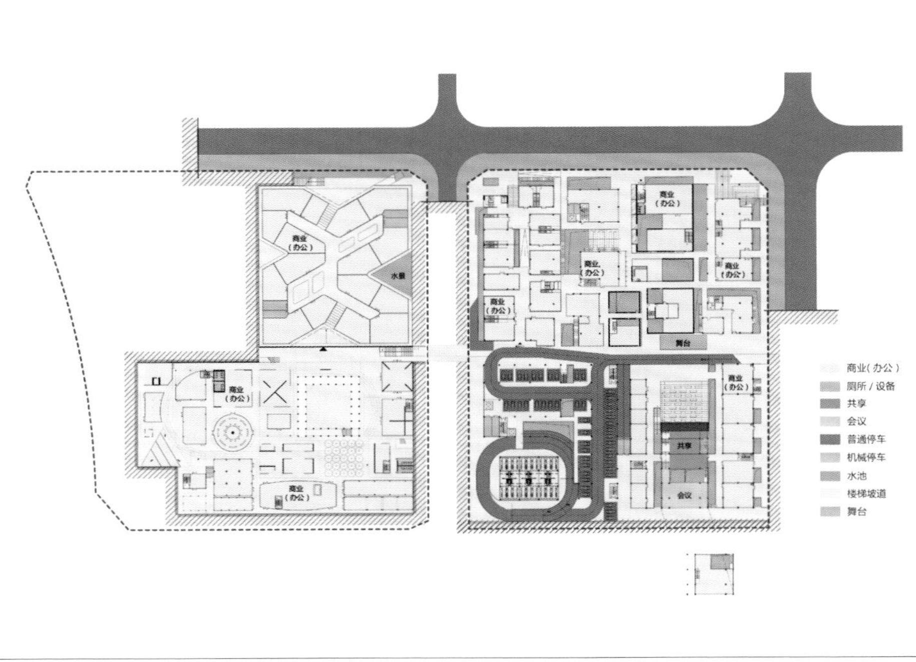

红线
退线
10.00 (46.3)
5.00 (41.3)
±0.00 (36.3)
-1.200 (35.1)
-6.200 (30.1)
-11.20 (25.1)
5000
5000
10000
1200
5000
11200
5000
人行道
商业（办公）
车库入口
走道
中庭
展览
商业（办公）
无人车
WC
商业（办公）
商业（办公
1:300

19
Design Commune

设计公社（A3+B4，A4+B2）
城市地下空间的集约化利用

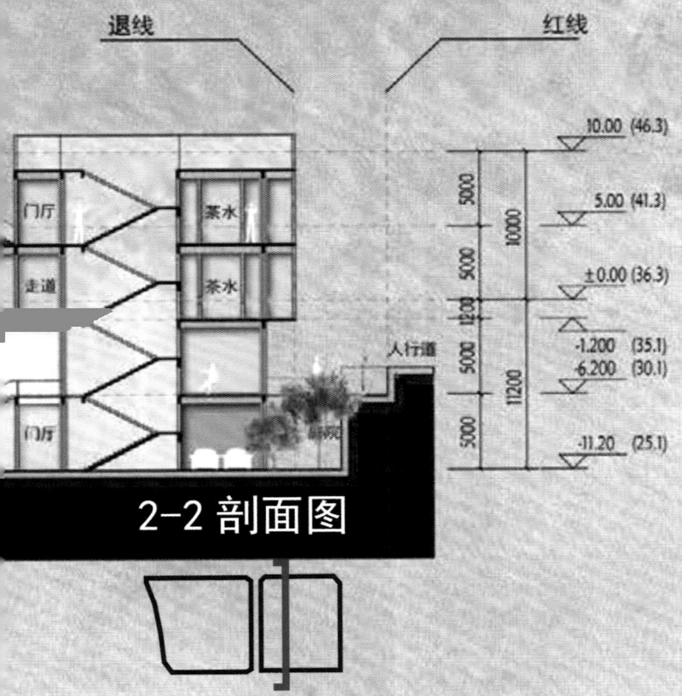

2-2 剖面图

总体城市设计和景观设计：都市实践
建筑设计：南沙原创建筑设计工作室；坊城建筑设计
项目地点：深圳
设计时间：2014 年 5 月—2016 年 8 月；
2014 年 7 月—2016 年 9 月
建筑总面积：7,992 ㎡；7,500 ㎡
功能：建筑设计

Overall Urban Design and Landscape Desig: URBANUS
Arch. design: Nansha Original DEsign ; FCHA
Location: Shenzhen
Design process: 2013.05—2016.08 ; 2014.07—2016.09
Site area: 7,992m^2; 7,500m^2
Program: Building

留仙洞万科云设计公社位于深圳南山区大沙河创新带，科技园北区，是留仙洞总部基地九街坊规划中的北绿廊地块。万科想借此机会在北绿廊打造一个集合产业链上下游的创意园区—设计公社。万科与项目总规划单位都市实践（URBANUS）对设计单位进行筛选及召集，最终确定在都市实践的指导下，通过集群设计的方式来完成 05-02，05-01 地块这个首开启动区的设计。该项目深圳板块的四支建筑集群团队包括了坊城设计、南沙原创、华艺和奥博能，其中 A3 和 B4 的建筑方案由南沙原创完成，A4 和 B2 的建筑方案由坊城建筑设计完成。

这是一个半命题式地下空间建筑设计项目，地面层必须作为公共绿地公园，地下空间有条件地提供给小型创意性企业作为办公和公共服务设施使用，希望形成一个别具一格的创意社区。

Design Commune is located in the Dashahe Innovation Zone in Nanshan District, Shenzhen, and in the North District of the Science and Technology Park. It is the North Green Corridor site in the planning of the 9th Neighborhood of the headquarters of Liuxiandong. Vanke would like to take this opportunity to create a creative park-design community in the North Green Corridor that integrates the upstream and downstream of the industry chain. Vanke and the project overall planning unit URBANUS selected and convened the design units, and finally determined that the design of the first start-up area of the 05-02, 05-01 plot should be completed through Cluster Design under the guidance of URBANUS. Four teams, i.e., FCHA, NODE, HUAYI, and URBANERGY, were designated for the architectural design of building clusters on Shenzhen plots, of which A3 and B4 were designed by NODE, A4 and B2 were designed by FCHA.

This is a semi-propositional experimental underground development. The ground level must be spared for public green, while the underground level may be used as creative office and public service facilities if possible and to establish a unique creative community.

1. A3+B4，A4+B2 地块与整体平面图底关系

1. Figure-Ground Relation of A3+B4,A4+B2 Block and Overall Plan

A3 地块俯视图

Top view of Plot A3

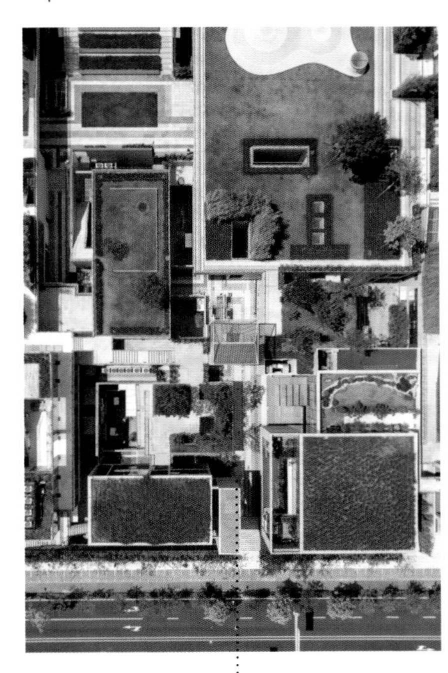

B4 地块俯视图

Top view of Plot B4

A4+B2 地块俯视图

Top view of Plot A4+B2

A3+B4 南沙原创建筑设计工作室

A4+B2 坊城建筑设计

A3+B4 Nansha Original DEsign

A4+B2 FCHA

B4

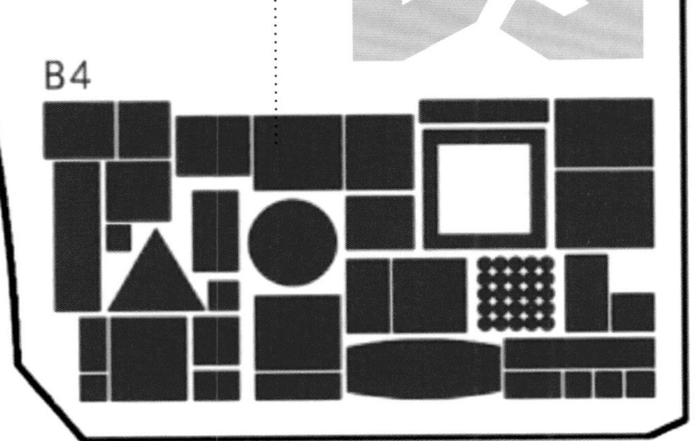

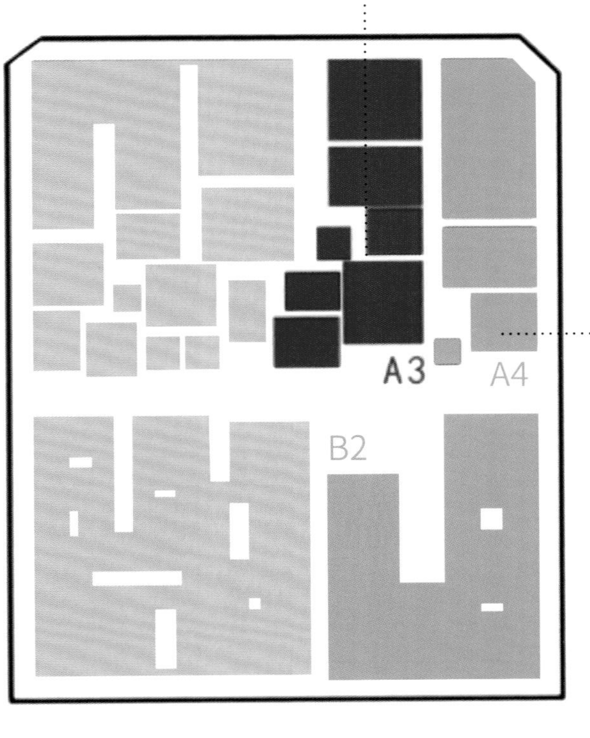

“人的一生很短暂，为自己工作，为自己喜欢的东西工作，这才是最重要的。”

关于建筑师

ABOUT THE ARCHITECT

陈泽涛
Chen Zetao

刘珩
DOREEN HENG LIU

坊城设计（FCHA）：坊城设计传达了我们对待城市、街坊、建筑的设计理念，既要从更广阔的城市分析来切入街坊及建筑设计，同时也要在建筑设计中体现街坊的邻里关系及生活，同时也表达了我们以建筑类型学这样一种研究方法对中国城市化进程的思考。

有关坊城建筑设计，详见 P22 页。

FCHA: a company that translates our concepts of Fang, CHeng and Architecture. We analyze architecture design with a sense of urban planning, and care for neighborhood and living style during the design process, producing research methods with architecture typology in China's urbanization process.

For details on FCHA, see page P22.

南沙原创，中国南方当下具有设计影响力并受到国内外广泛关注的独立建筑设计事务所之一，由刘珩于 2004 年在南沙创建，2009 年底设立深圳办公室，目前有近二十个建筑师组成的专业建筑设计团队。

多年来坚持对建筑基本问题的研究和实践，追求在严谨务实基础上的创新；同时也在建筑理念上探索其自身逻辑对跨领域的开放性和兼容性，并以此作为设计的出发点。近年来尤其关注在城市公共空间及城市更新设计，文化艺术教育建筑以及公共基础设施领域的跨界研究和建筑实践，通过领域间的互动和激发，保持自身在建筑领域的前瞻性和实验性。

NODE stands for Nansha Original DEsign (or NO DEsign) and was established in 2004. It is a small and high quality architectural practice in the Pearl River Delta region, which grew out of a series of projects associated and completed with the Fok Foundation of Hong Kong, and has extended its practice geographically outward. Founded and led by Principal Architect, Doreen Heng LIU, NODE currently consists of 15 architects and designers from China and other countries, and has completed and is implementing architectural projects in China. For years, NODE, with Nansha as her local base and Hong Kong as her international window, also established a studio in Shenzhen in late 2009, has been conducting a diversity of architectural, art and design practices in the PRD region. NODE here is understood as the point in which different vectors – possibilities, tendencies, and events – intersect. It is neither a point of departure or of arrival. It is not a fixed entity. It is determined by ever-changing fluxes and dynamics.

1.2013 深圳双年展浮法玻璃厂改造
2—5. 设计公社实景图

1. Floating Entrance,Fufa Glass Factory Renovation
2—5. Design Commune Scene

梯城——留仙洞北绿廊云城 A3 地块

A3 地块的设计，首先需要遵循城市设计团队所规定的集群设计的共同游戏规则：城市肌理、公园覆土和建筑限高的要求；在这样的先决条件之下，各自团队如何做出自己的特色？其次，面对南北地块的地面落差、地下空间的通风采光消防等实际问题，地下空间如何创造一处适合创意人群工作和交流的场所？设计的挑战是多维度的。

由于限高，大部分建筑的使用空间在地平线之下，为增加其通风采光，以独立院落作为建筑原型，每个建筑由 2 层空间 / 楼梯 / 花园三个元素构成，虽然有共同的元素，不同的穿插与重组，以及所处的城市设计位置和相邻关系的差异，每个院落又各有特点，相互作用又相互独立，可分可合；组合在一起，虽然都置于地下，但是有天有地、有风有水。同时，高低错落、迷宫式的路径也让这个日常场所有了别样的体验感。

桌景——留仙洞北绿廊云城 B4 地块

B4 地块，一个占地超过 7000 平米的地下，地面平层让人很容易联想到一个平整的桌面，如果这是一组不同形式的桌面堆在一起呢？方案采用了化整为零的手法，整体而言，它还是一个平面，但有趣的是，如果你钻到桌底看，看到的却是另一道风景：不同形式的桌子，会由不同形式的结构组成，之间可能会形成意想不到的空间，也就形成了空间之间的对话，而且桌子之间的缝隙就有了自然采光和自然通风的可能。可以想象每一个地下建筑单体都是一张抽象意义上的桌子：桌面是城市绿地，桌底就是地下层，建筑实际的使用空间就是在结构基础上设计组合的。一个个独立、简单、不一样的“桌子”共同组成一个近 4800 平方米地上与地下的双重景观。

确定了以结构的多样性塑造空间关系后，在 3M 为基本网格模数控制功能尺寸的基础上，以罗马 NOLLY MAP 的城市灰空间和岭南地域应对气候的建筑灰空间 – 公共街道与建筑之间的虚实肌理关系为灵感，将篮球场、森林广场、看台、咖啡厅等公共功能穿插于办公空间之间，形成了层次丰富的体验空间及路径。

Terraced-City——Liuxiandong Vanke Green Corridor North Plot A3

The design of Plot A3 should, first of all, follow the acknowledged rules set for the design of building clusters by the urban design teams, i.e., the requirements of urban fabric, park earth cover and maximum building height; under such circumstances, how could each team bring out its own strengths? Besides, with height difference between ground level of north and south plots, and the challenges of underground spaces including ventilation, daylighting and fire protection etc., how could a place for work and interaction be created for creative people? There were multiple challenges to overcome in design.

Due to height limit, most building spaces are below ground level. In order to improve ventilation and daylighting, independent courtyard was adopted as architectural prototype, each consisting of three elements, i.e., two floors of spaces/stairs/garden. Despite these common elements, there are also varied ways of alternation and re-assembly, and different urban design locations and neighboring relations. As a result, these courtyards, each with its own features, are mutually interactive yet independent. They can be either separated or combined; although being placed below grade, they also enjoy generous view and pleasant ventilation and landscape just like those at grade. The staggered labyrinth routes also enable a special experience in this ordinary place.

Table-Landscape——Liuxiandong Vanke Green Corridor North Plot B4

The flat level at grade is easily reminiscent of a smooth tabletop. What if it's composed of several tabletops in varied forms? Inspired by the idea of breaking up the whole into parts, we create a tabletop that appears as an integral whole but is actually propped up by a number of tables in different structural forms. Unexpected spaces are thus created to dialogue with each other, while the gaps between tables allow for the possibility of natural daylight and ventilation. Each underground singular building may be conceived as a table in abstract sense, with the tabletop as urban green and the space underneath it as below-grade levels. The occupied spaces of the buildings are designed and combined on top of the structure. Those independent, simple and differentiated tables together form a nearly 4,800 sqm of double-level landscape at and below grade.

After defining the spatial relations via structural diversity, we employ a basic grid module of 3M for control of functional dimension, and intersperses office spaces with such public functions as basketball court, forest square, stand, and coffee shop to offer experience spaces and routes of diverse hierarchies. Such a concept is inspired by the void-solid relations between public street and buildings as shown in the urban gray space of NOLLY MAP in Roma and the climatically responsive gray building spaces in Lingnan region.

1. 七座建筑形态概览
2. 公共空间景观
3. 楼梯细部

1. The Forms of Seven Architecture Overview
2. Landscape of Public Space
3. Staircase Details

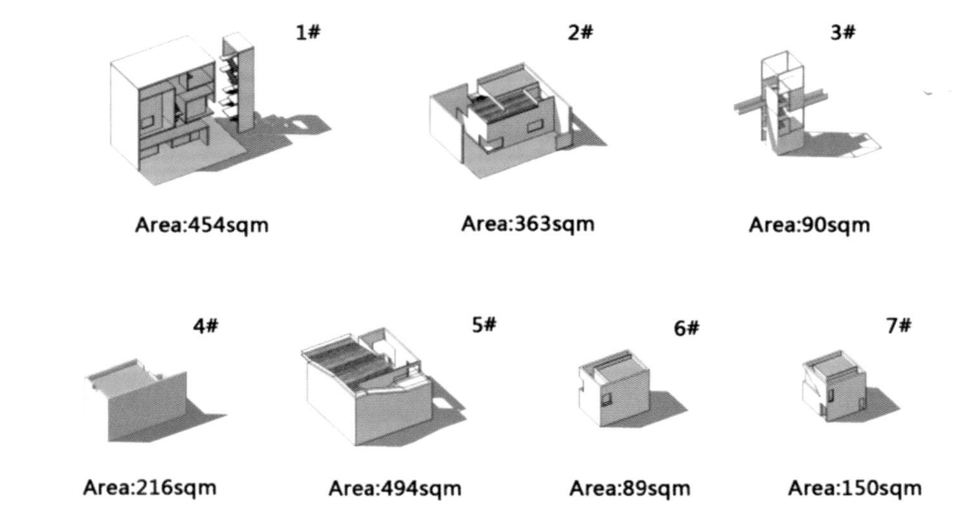

A4+B2 地块

A4+B2 地块占地面积 7500 平方米，地块呈刀把形，长宽为 135 米 × 50 米，场地南高北低呈缓坡趋势，高差为 3.7 米。在城市设计导则的指引下，设计通过大绿化坡道将首层与负一层连接，划分连续的办公单元，将北向区域高起作为独立建筑单元，增加自然采光通风洞口。在保证屋顶公共绿地的连续性与提供地下办公空间良好环境之间寻找平衡点。在功能划分上，将空间分别设计为办公单元、共享空间、小型报告厅、共享办公台地、独立办公区、创客工坊、采光天井、交通楼梯、交流平台、自然岩石墙面及通风廊道。

材料由集群设计师多方经过商讨后决定统一使用清水混凝土、钢、玻璃等自然材料来进行建造、明确营造一个语言统一、空间丰富、产品多样的园区。

A4+B2 plot

A4+B2 plots cover an area of 7,500 square meters. The plots are knife-shaped, the length and width are 135 meters × 50 meters. The site is high in the south and low in the north, showing a gentle slope with a height difference of 3.7 meters. Under the guidance of the urban design guidelines, the design connects the ground floor and the basement level one floor through large green slopes, the plot was divided into continuous office units, and raises the northward area as an independent building unit to increase natural lighting ventilation openings. A balance is maintained between ensuring the continuity of the public green space on the roof and providing a good environment for underground office space.

As for materials, after many discussions, the cluster designers decided to use fair-faced concrete, steel, glass and other natural materials for construction. It has been made clear that a park with unified language, rich space and diverse products should be created.

1. 剖面业态示意图
2. 项目鸟瞰图，材料与自然相融合

1. Section with Program Diagram
2. Birdeye View, Materials Corresponded to the Nature

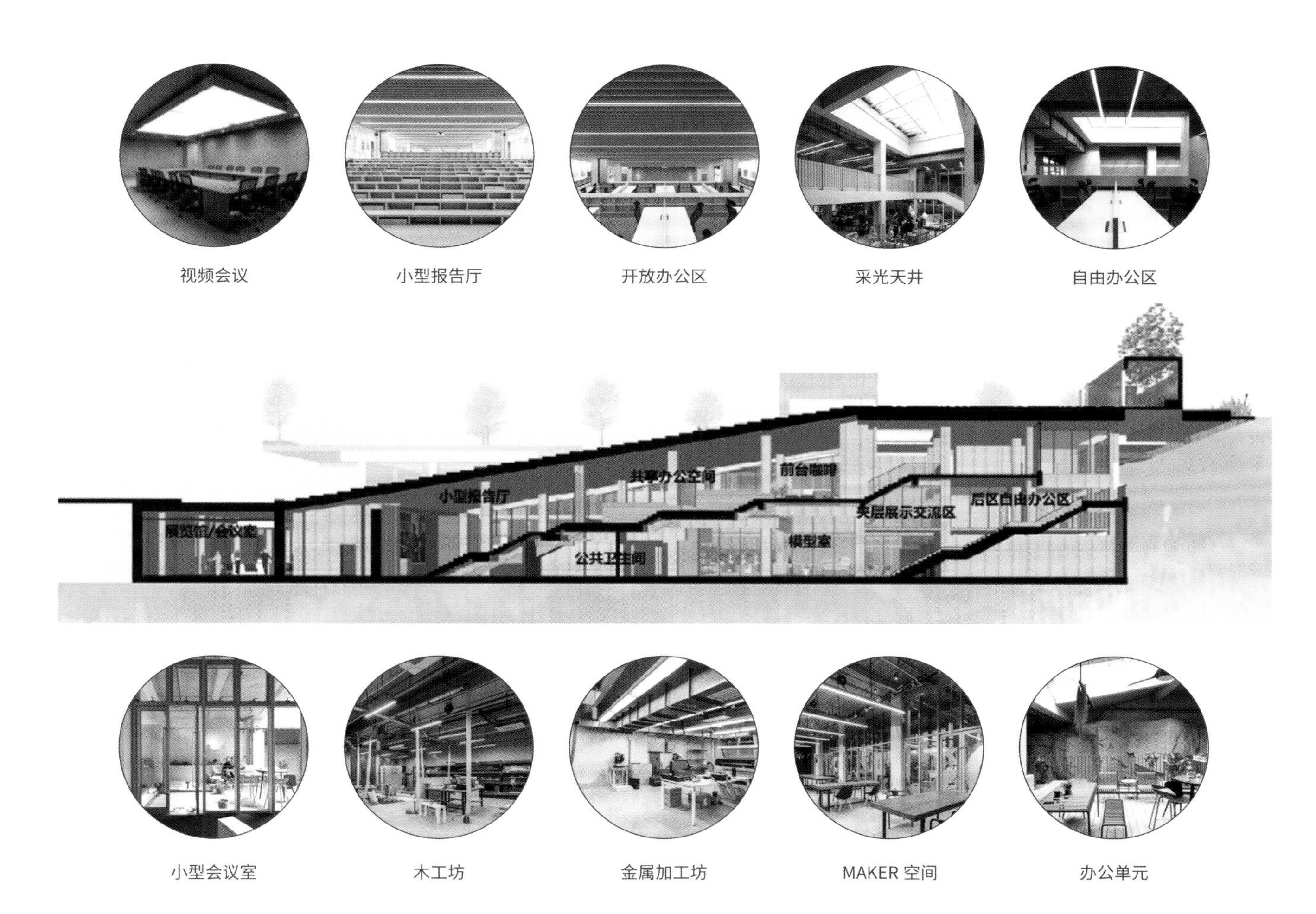

1. 场地中央的楼梯
2. 层次丰富的体验空间及路径
3. 木制阶梯组成的小型报告厅
4. 自然岩石墙面及通风廊道

1. The Staircase in the Middle of the Ground
2.Intersperses Office Experience Spaces and Routes of Diverse Hierarchies
3. The Small Lecture Hall Composed of Wooden Steps
4. Natural Stone Wall and Ventilated Corridor

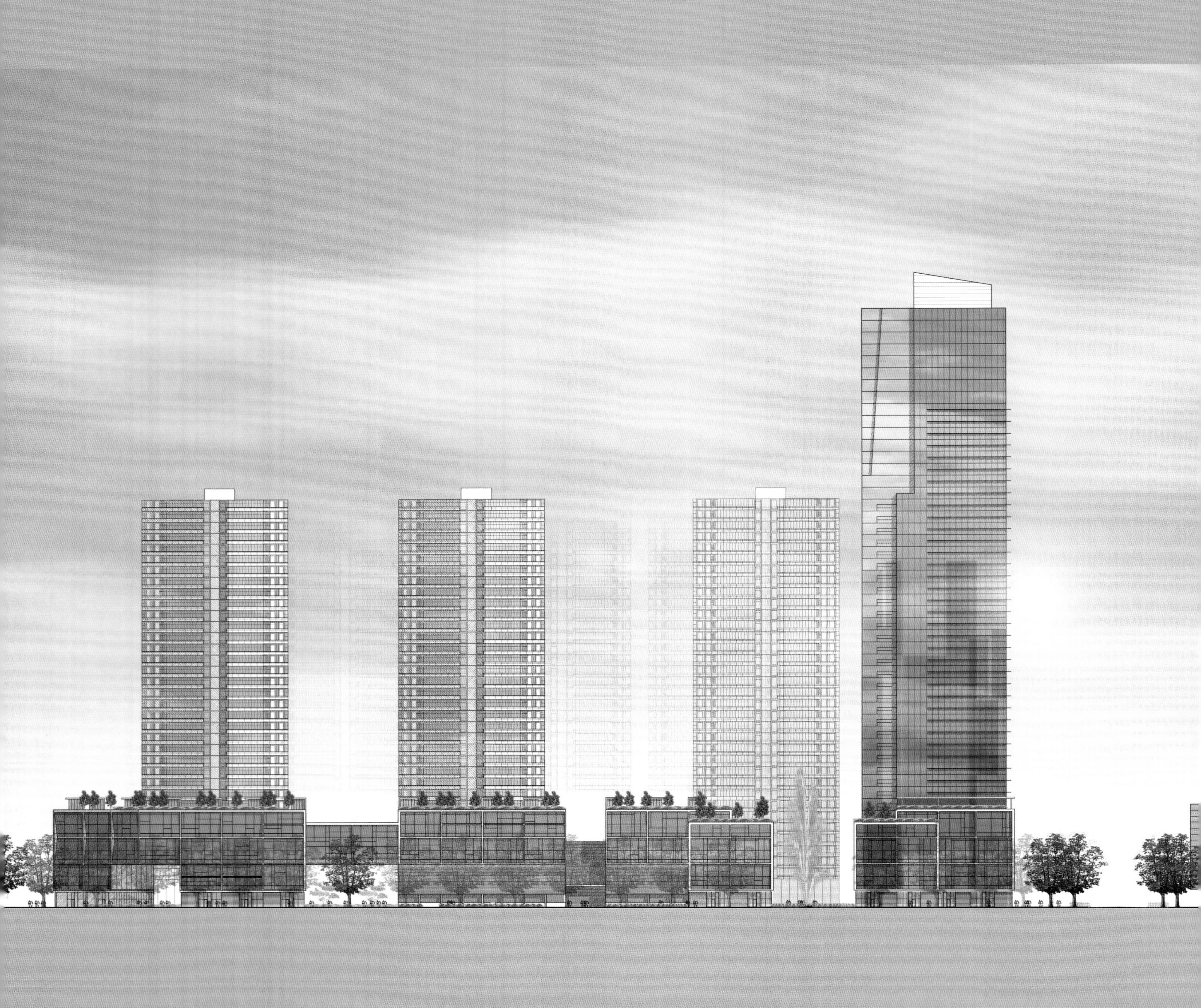

20 Vanke Longgang Center Masterplan

万科龙岗中心规划方案研究

多元化城市的挑战，来自世界的答卷

龙岗中心城项目位于中国深圳市龙岗区未来的商务中心区（CBD）核心部分，距市中心罗湖 / 香港边界仅 45 分钟轻轨公交车程，为深圳市直属六个大区之一，经重新规划设计后，项目地段与 CBD 区内其他地段一道将成为有服务业、商业、金融和政府机构的多功能的重要城市中心，区域内的人口将超过 350，000 人。万科房地产开发有限公司作为该地区开发的重要成员之一，为了更好地配合龙岗区城市设计总规划，高质量地完成龙岗中心城项目的策划与开发，组织了一次国际城市设计咨询活动。在小规模邀请国际知名事务所参与咨询活动后，甄选了十个各具特色的项目作品，通过他们拥有的专业班底和丰富的国际经验，再通过对本项目的研究或对现有规范的再解释，最终为本规划地段制定出一个有个性和特征的品牌方案。

本次书中精选出其中三个富有创造性、实用性和可实施性的案例：一个来自荷兰著名建筑事务所麦肯诺，由世界知名女性建筑师荷本女士领衔设计；一个来自以“温哥华式建筑”闻名的郑景明事务所；另一个来自澳大利亚的登顿・科克・马歇尔事务所。

建筑设计：麦肯诺，郑景明，登顿・科克・马歇尔

项目地点：深圳市龙岗区

功能：总体规划

Longgang City Center locates at the core of the future CBD of Longgang District in Shenzhen City, China. Close to Luohu and only 45 min bus ride to Hongkong periphery, Longgang is one of the six major districts of the city, and the City Center, together with the rest of the Longgang CBD areas after the latest master plan, will become a vital core of the district that combines multiple functions such as services, business, economic and government departments. The number of residents would exceed 350,000. Vanke, as one of the most significant developer of the Longgang District, had hosted a small-scale international urban planning consultation forum. 10 featured schemes were selected from international renowned design studios, with professional and experienced team members planning and researching the most appropriate and characterized master plan.

Three elite schemes, all of which with creative plan and practicality were included in this book: from Dutch design studio Meacanoo, James KM Cheng studio from Canada, and DKM from Australia.

Arch. design: Mecanoo Architects, James KM Cheng Architects, Denton Corker Marshall

Location: Shenzhen Longgang District

Program: Masterplan

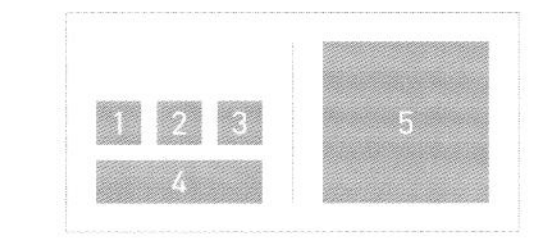

1. 鸟瞰图
2. 模型示意图
3. 夜景模型示意
4. 立面图
5. 标志性建筑效果图

1. Birdeye View
2. Model Photo
3. Nightview Photo
4. Elevations
5. Iconic Building Render

1. 来自郁金香之国 —— 荷兰设计
Dutch Design—From the Hometown of Tulips

荷兰麦肯诺事务所方案
MEACANOO

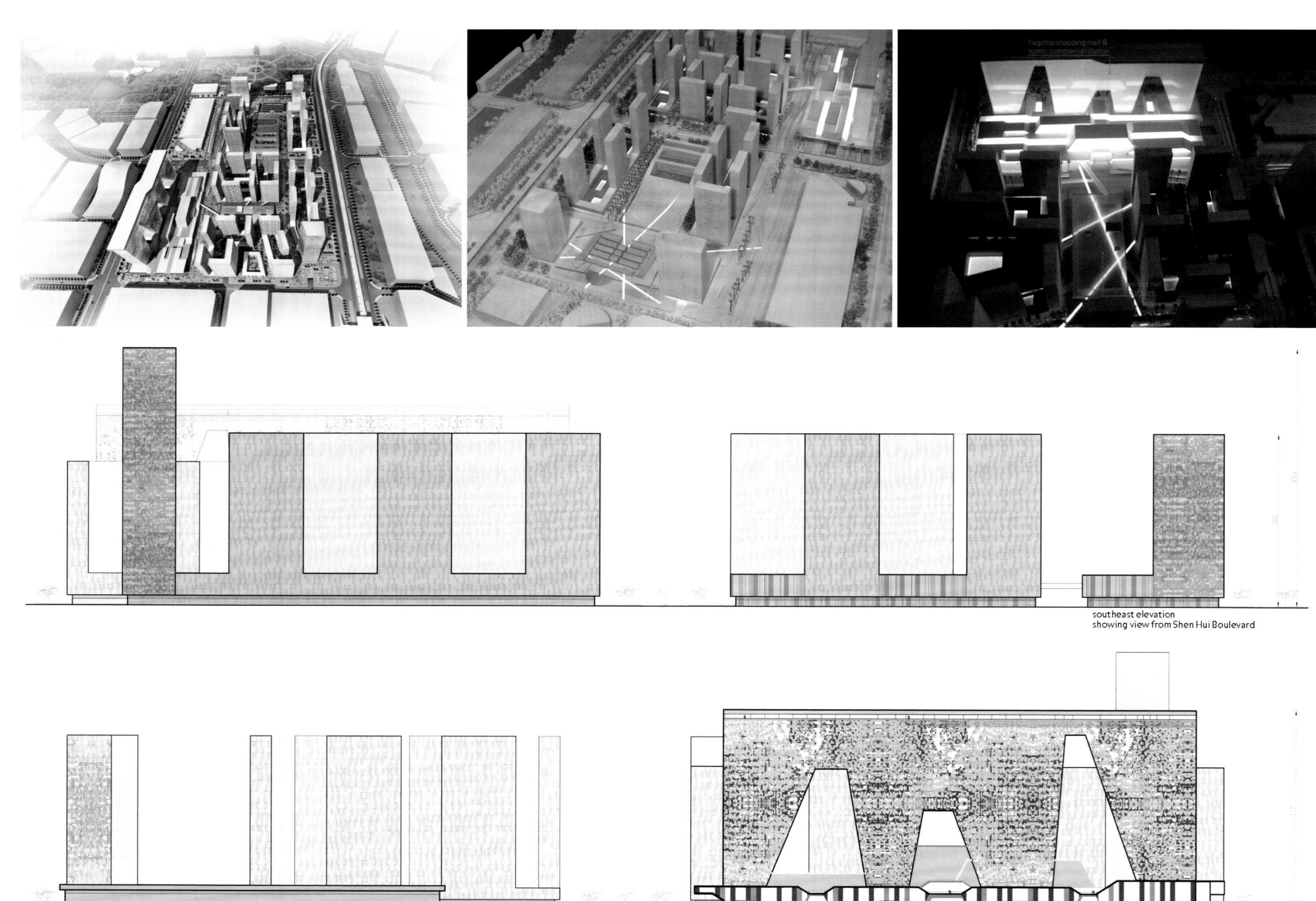

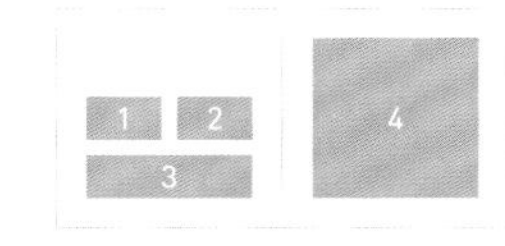

1. 总规划图
2. 模型图
3. 立面图
4. 总规划模型鸟瞰图

1. Masterplan
2. Model
3. Elevation
4. Model Birdeye View

2. 优雅飘逸 —— 加拿大的“温哥华主义”
Elegant and Graceful — “Vancouver Style” from Canada

加拿大郑景明建筑师事务所方案
James KM Cheng Architects

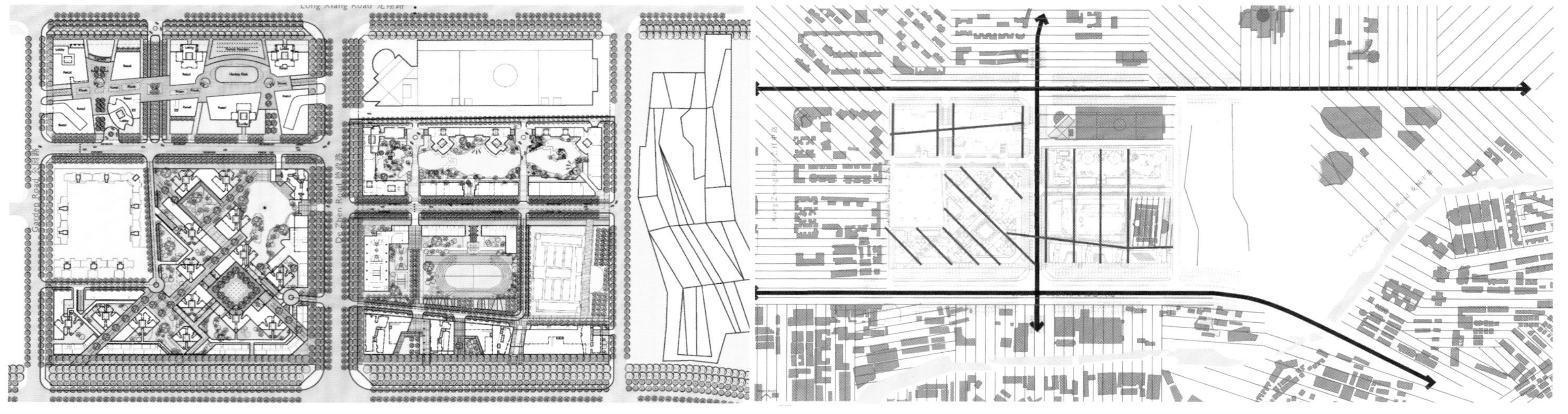

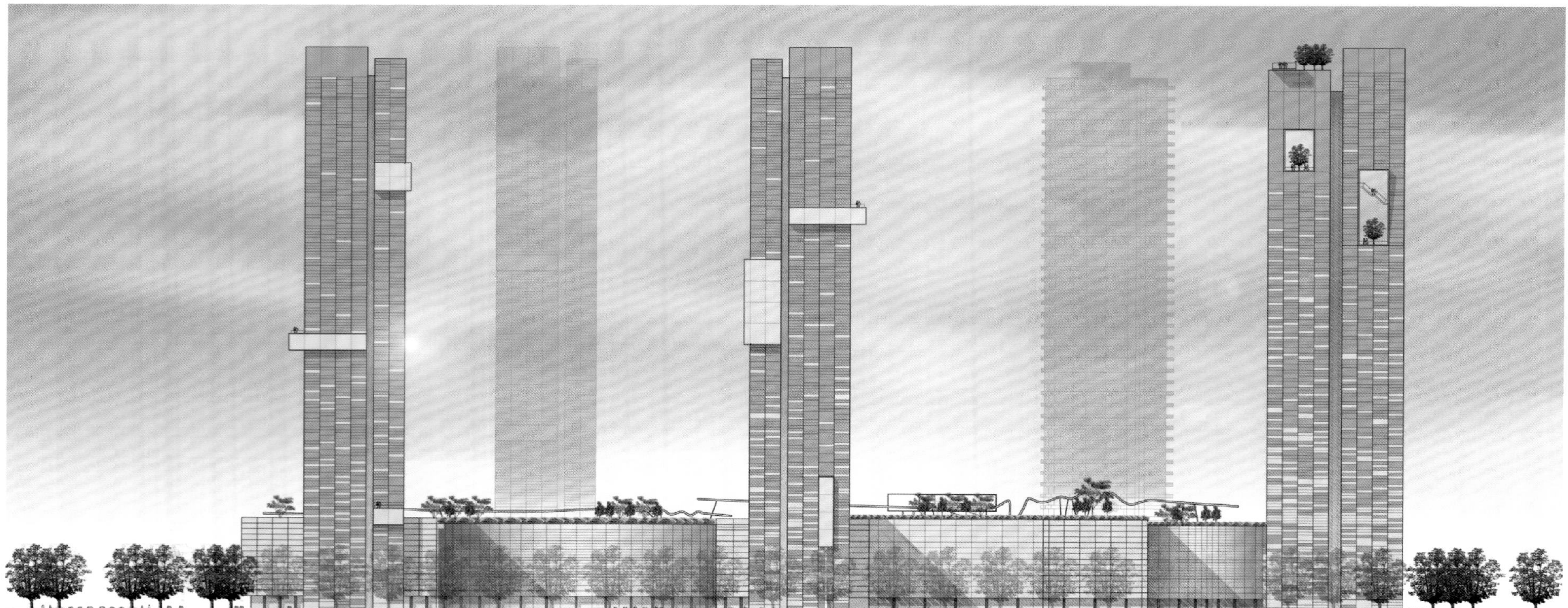

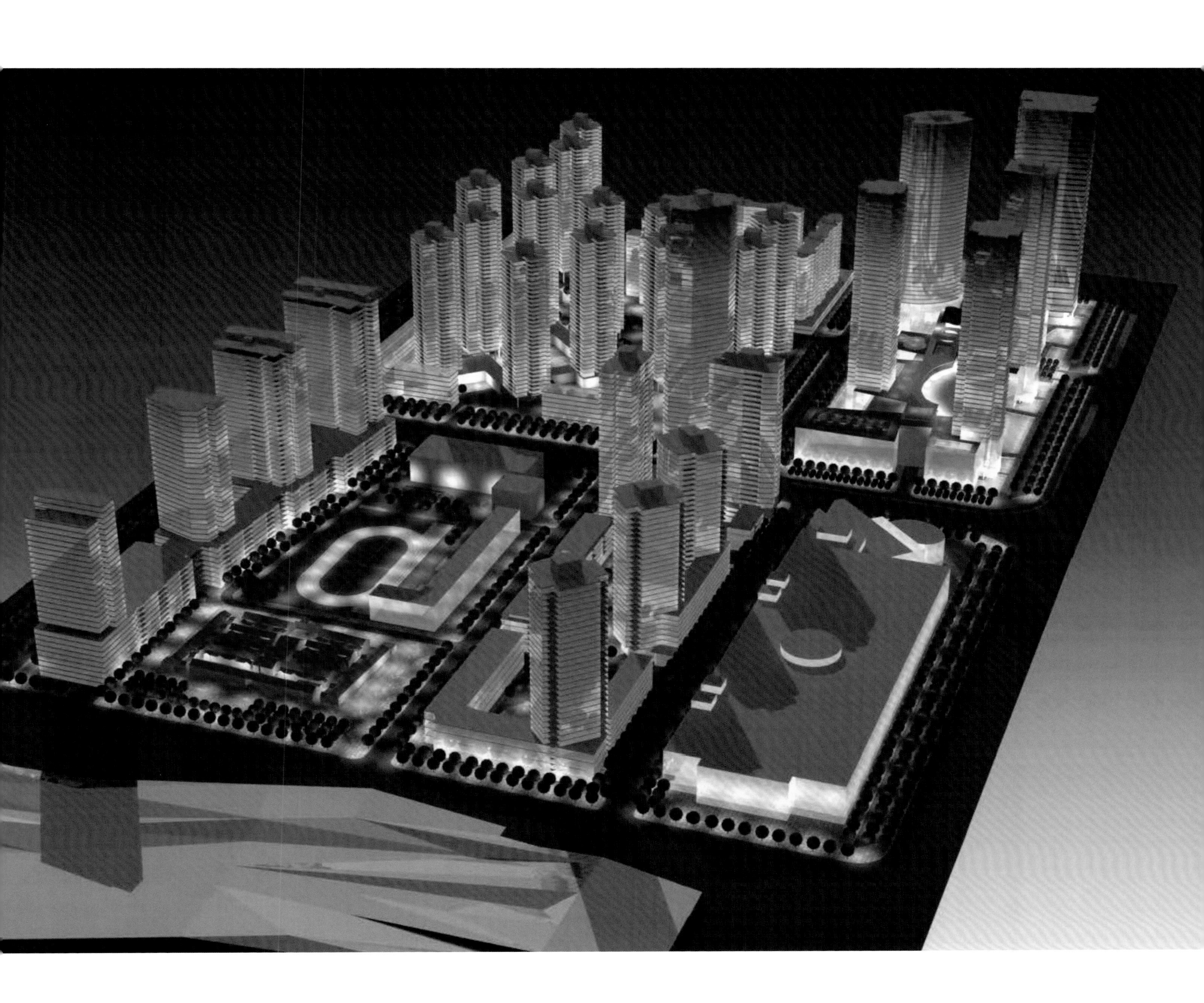

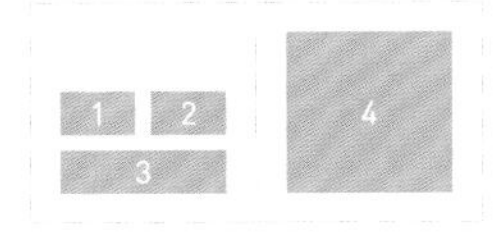

1. 总平面图
2. 街境渲染图
3. 鸟瞰图
4. 鸟瞰图

1. Masterplan
2. Streetscape Render
3. Birdeye View
4. Birdeye Vew

3. 太平洋南岸的白色之风 —— 澳大利亚的色彩
Color of Australia — Breeze from the South Pacafic

澳大利亚登顿 • 科克 • 马歇尔事务所方案
Danton Corker Marshall

21
Highline Park

高线公园
与山共舞

建筑设计：集合设计
项目地点：中国苏州
设计时间：2017 年
功能：商业与公园

Arch. design: One Design
Location: Suzhoug, China
Design process: 2017
Program: Commercial and Park

苏州，如同一部有迹可寻的最古老的浸没式戏剧，是本土化与全球化的现象级文化标本。昆曲 + 园林是浸没式戏剧之鼻祖，昆曲是流动的园林，园林是凝固的昆曲。苏州高新区的高线公园则是将戏剧、园林和社交相互交织的城市空间剧本。

高线公园是一种丰富的时间结构，串联起各种关于都市和自然的体验，可以和这个城市一起衍化生长。当城市空间剧本和时间结构相互成长，相互依存，在苏州这座古老美丽的城市之上激发出独辟的境界。

Suzhou is a city like an ancient immersive theatre that still active on stage nowadays, it is a live cultural specimen of localization and globalization. The southern opera style, Kun Opera, together with the traditional Suzhou garden, have provided probably the earliest "immersive theatre" experience. Kun Opera is a fluid garden with melody, and the garden became a line of still melody of the Opera. The SND Highline Park Project has intertwined urban life scripts into the elements of theatrical experience, traditional Suzhou garden style and community lives.

The Highline Park in Suzhou, as a structure that contains a timeline concept, should not provide connection between points in the city and nature, but also grow and mature with the city itself. The spatial script and timeline structure are woven and interdependent with each other in the city, The park will be a unique space in such a beautiful ancient city.

1. 总规划图
2. 轴测图
3. 鸟瞰图
4. 截面图

1. Masterplan
2. Axonometric
3. Birdeye View
4. Section

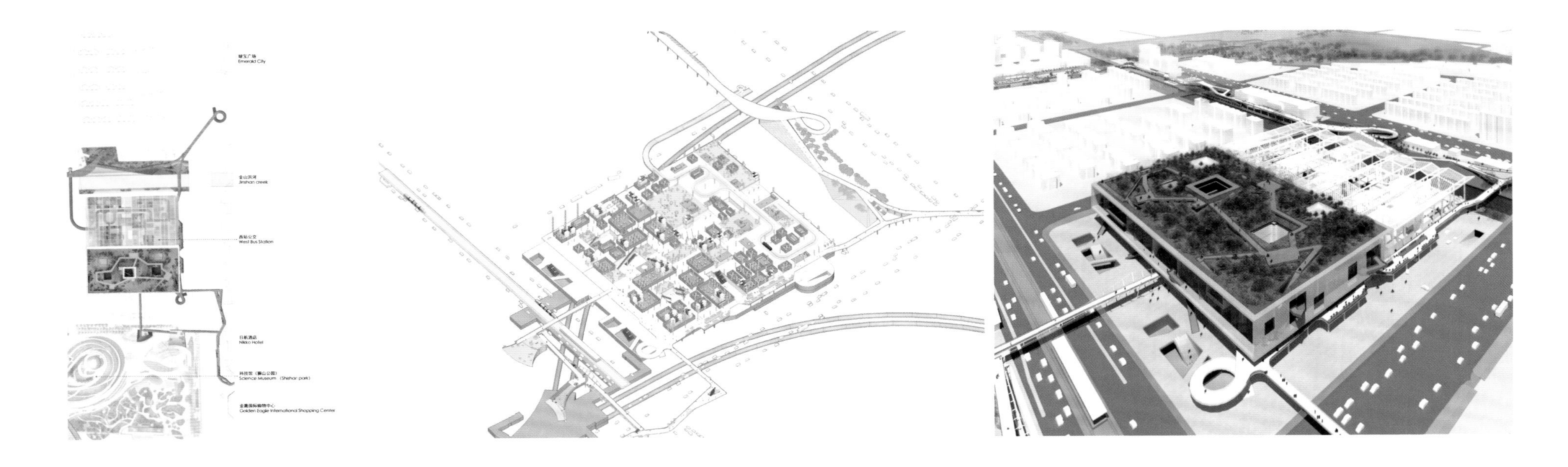

“设计作为一种为了解决问题的形式化、策略性的手段，是可以跨越尺度和领域而在思想层面上自由运用的。”

关于建筑师

卜冰在 2014 年发表于《世界建筑》的一篇文章中提道，他 1991—1996 年在清华求学时并没有非常快乐，而是常有各种怀疑和焦虑，在信息相对的匮乏中读着“后现代主义”的理论，听着“民族符号”、“古都风貌”和“经济实用的前提下尽可能美观”的声音。好在各位老师提供了多样的意见和帮助，加深了他探求真相、自主选择的欲望，在慢慢培养自己价值观的同时保持心态的开放。后来他到耶鲁深造，于 2000 年加入马达思班建筑设计事务所，并于三年后创立集合设计。其作品包括南京红枫科技园、宁波五龙潭山庄、慈溪 CBD 和上海朱家角西镇的城市设计，以及在中美多地展出的云屋装置。他还多次在国内外策展，并于 2011—2014 年担任美国圣路易斯华盛顿大学与同济大学的研究生联合城市设计课程导师，于 2015 年起担任美国雪城大学的亚洲城市研究与设计课程导师。

ABOUT THE ARCHITECT

Bu Bing had mentioned in an article in *World Architecture Magazine* published in 2014 that he was constantly worried and feeling anxious during the study period in Tsinghua University back in 1991 to 1996. That was the time where informations were relatively difficult to obtain, he was reading books on Post-modernism and bombarded with the voices of "national symbols", "traditional cityscapes" and "economical-and-practical-based-aesthetics". Luckily his mentors were helpful enough to guide him into a path of autonomous learning and critical thinking. Bu Bing gradually found his own opinion in an open-minded journey of architecture studies in Yale and early career in MADA s.p.a.m from 2000. Three years later he founded ONE Design, and completed projects such as Nanjing Hongfeng Technology Park, Ningbo Wulongtan Mountain Villa, Urban planning for Cixi CBD and Shanghai Zhujiajiao West, his installation "Cloud House" had toured multiple exhibitions in China and US. Meanwhile, Bu Bing is curator of many domestic and international exhibitions, he previously taught at Washington University in St. Louis and Tongji University in International urban design masterclasses, since 2015 he works as a tutor in Syracuse University on Asian City Studies.

卜冰
Bu Bing

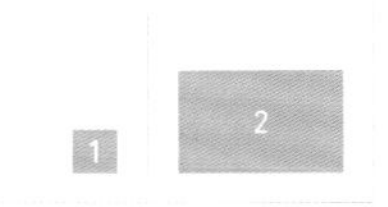

1. 南京红枫科创园
2. 鸟瞰渲染图

1. Nanjing Hongfeng Tech Park
2. Birdeye View Render

PART G Heritage Preservation And Upgrade
历史保护和更新

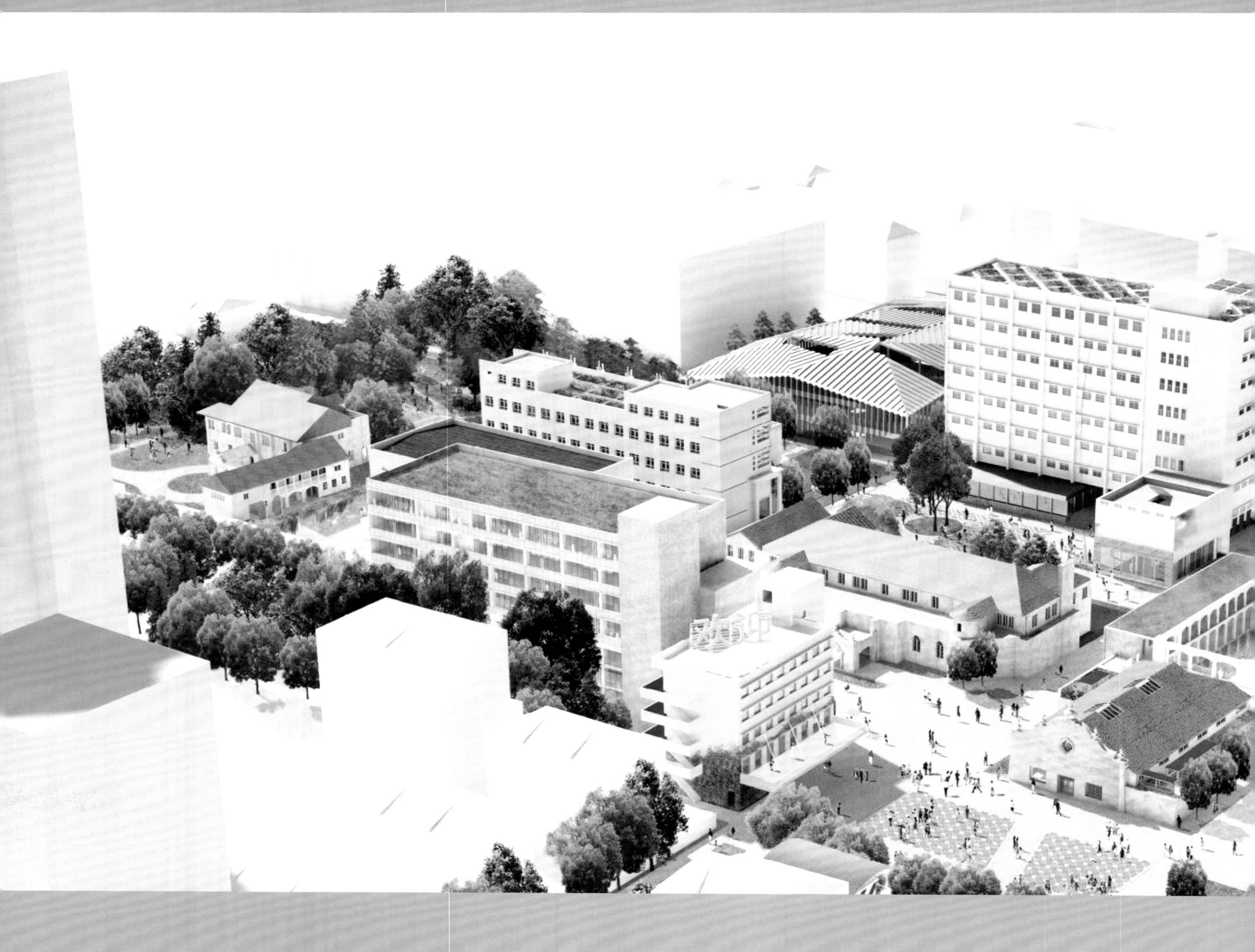

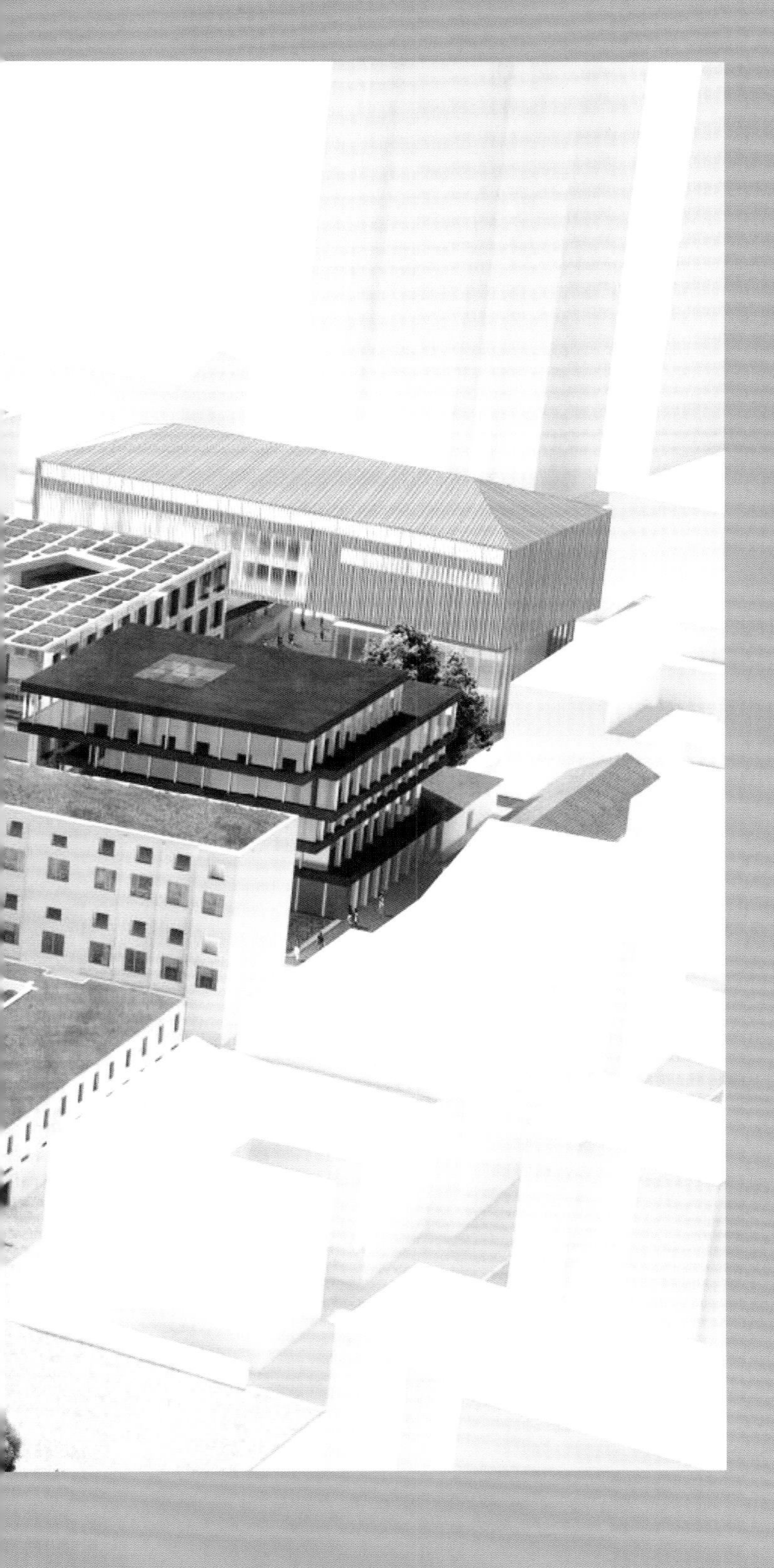

22 Shanghai Columbia Park

上生新所

时间的万花筒

建筑设计：大都会建筑事务所

项目地点：中国上海

建筑总面积：48,000 m²

功能：混合

Arch. design: OMA（Office For Metropolitan Architecture）

Location: Shanghai, China

Total area: 48,000m²

Program: Mixed Use

大都会建筑事务所（OMA）与当地 ECADI 建筑事务所以及 West 8 景观设计公司共同合作，为在上海市中心的哥伦比亚公园做出综合整体规划。项目所在位置是一个历史悠久的区域，至今还保留着殖民时代的遗迹。由建筑师埃利奥特 • 哈扎德（Elliott Hazzard）设计的工业厂房和 20 世纪 20 年代的乡村俱乐部等，这些建筑将通过此次总体规划进行更新和改造，并将哥伦比亚公园重新塑造成上海最著名的公共空间之一。

在这片 4.7 公顷土地上现有的建筑中就包括哥伦比亚乡村俱乐部，是由哈扎德为战前的美国精英社团所设计，里面有会所、健身房和室外游泳池。 酒店内还设有一座由匈牙利建筑师邬达克（László Hudec）于 1930 年设计的别墅、三座西班牙殖民复兴风格的纪念碑和建于 20 世纪 50 年代至 2000 年期间的近 40 个建筑，各有用途也各具特色。

在 1952 年，上海生物制品研究所（SIOBP）接管了该地区，并将广场改造成了拥有办公楼、生产设备、仓库和实验室的研究园。后来自从 SIOBP 终止了在园内的各项活动，这里就处于休眠状态。这个潜力无穷的建筑正等待着被重新发现、赋予新生的一天。

在 OMA 的总体规划中，主要策略为重建原有的建筑结构，同时还要建设三座新建筑物，以此可以进一步扩展空间，并提供原有建筑无法提供的新功能。West 8 公司将负责创意办公空间、酒店、活动场所、文化场所、零售场所、餐厅和户外空间的设计，这些空间将分布在整个项目中，并营造出一个充满活力的体验场所。两个新的出入口将穿过邻近社区的主要通道，这也极大地改善了建筑的可及性。

OMA, in collaboration with local architects ECADI and landscape architects West 8, has developed a new mixed-use masterplan for Columbia Circle in the center of Shanghai. Layered with rich history, the site contains preserved colonial monuments, former industrial buildings and 1920's country club buildings by architect Elliott Hazzard – these elements will be renewed and transformed by the master plan to return Columbia Circle into one of Shanghai's most prominent public spaces.

Among the existing buildings on the 4.7 hectare site is the Columbia Country Club, designed by Hazzard for the prewar American elite society and including a clubhouse, gym and outdoor pool. Also on site are a villa designed by Hungarian architect László Hudec in 1930, three Spanish Colonial Revival Style monuments and nearly 40 additional structures built between 1950 and 2000, creating an eclectic mix of program and styles.

In 1952, Shanghai Institute of Biological Products took over the area, turning the square into a research campus with offices, production facilities, warehouses and laboratories. Since SIOBP ceased activities on the campus, the site has laid dormant, waiting to be rediscovered and repurposed with a project deserving of its architectural potential.

OMA's strategy will renew the existing structures as well as introduce three new buildings into the site, extending the repertoire of spaces even further and offering new programmatic possibilities the current historical buildings are unable to offer. Creative office space, a hotel, event spaces, cultural venues, retail spaces, restaurants and outdoor spaces designed by West 8 will be distributed throughout, creating an enlivened pedestrian experience. While two new access pointswill punch through to the main circulation arteries of the neighborhood, which improving accessibility to the site.

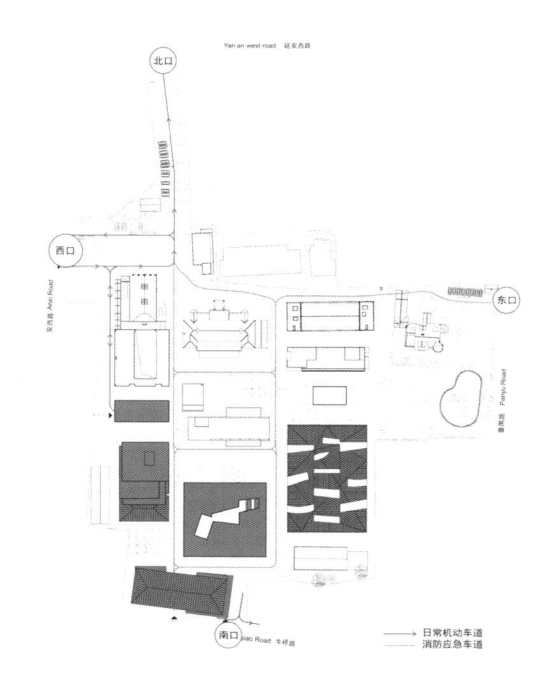

1. 交通流线图
2. 总规划图
3. 孙科别墅侧立面
4. 孙科别墅

1. Circulation Diagram
2. Masterplan
3. Side Elevation of Sun Ke Resident
4. Sun Ke Resident

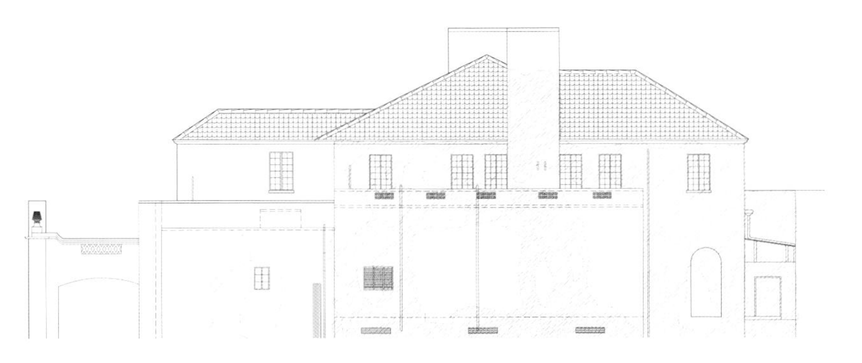

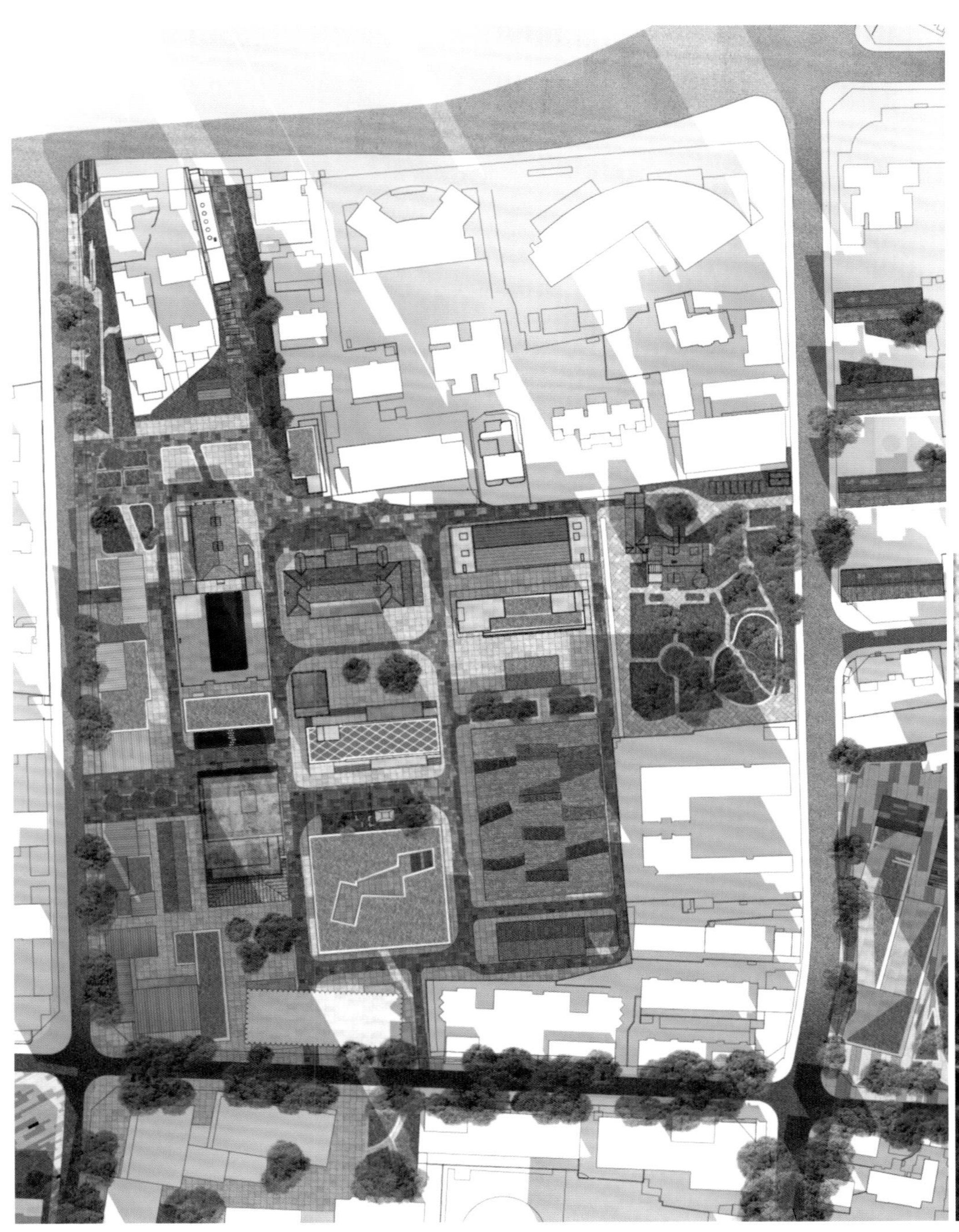

哥伦比亚乡村俱乐部（原哥伦比亚总会）始建于 1924 年，为上海市第三批优秀历史建筑，保护类别为三类保护。哥伦比亚总会是 20 世纪二三十年代上海西区侨民居住区——哥伦比亚住宅圈（Columbia Circle）的起始点和重要构成，建于约 1925 年，是近代上海外侨居住生活的重要例证。 哥伦比亚总会也是美国建筑师哈扎德在上海开业初期重要作品，其后哈扎德洋行成为在上海近代建筑史中留下重要作品的设计机构。哥伦比亚总会是给近代旅居上海的外国侨民留下美好记忆的娱乐场所，也是二战期间作为集中营时黑暗记忆的场所，部分侨民回忆录对这段历史有所记述，成为这些侨民及其后代的集体记忆。原哥伦比亚总会附属的游泳池及健身房，是哥伦比亚总会的组成部分，与总会主体建筑建造于同一时期。这两处建筑未列入保护要求，拟作为新增保留建筑进行保护再利用。

The Columbia Country Club, from which the project derives its name, was built in the 1920s during Shanghai's grand epoch, and it is classified as Class III historical buildings in Shanghai for preservation. The club's original buildings were designed in 1924 by American architect Elliott Hazzard for the American elite society, in 1952 the Shanghai Institute of Biological Products took over the site and the original buildings, and gradually developed a research campus with production facilities, offices, warehouses and laboratories. The Columbia Circle also includes former clubhouse, a gym and an outdoor pool, and they all belonged to part of the glorious memories of the early immigrants, as well as part of the dark memories of the concentration camp function during World War II. The gym and the pool was built under the same era but not classified as historical buildings, they are to be regenerated for modern purposes.

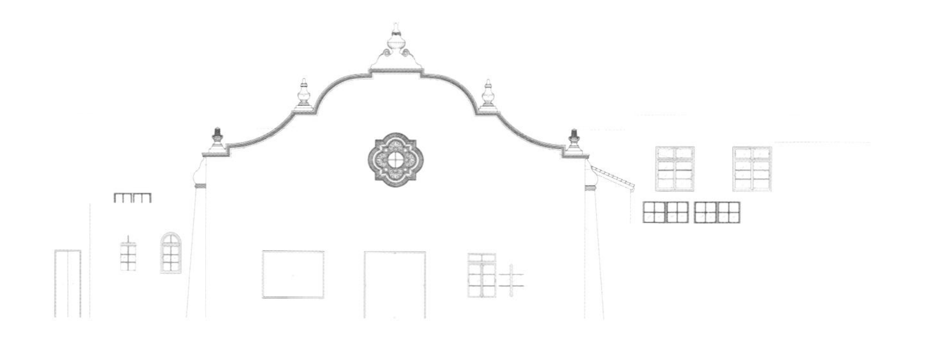

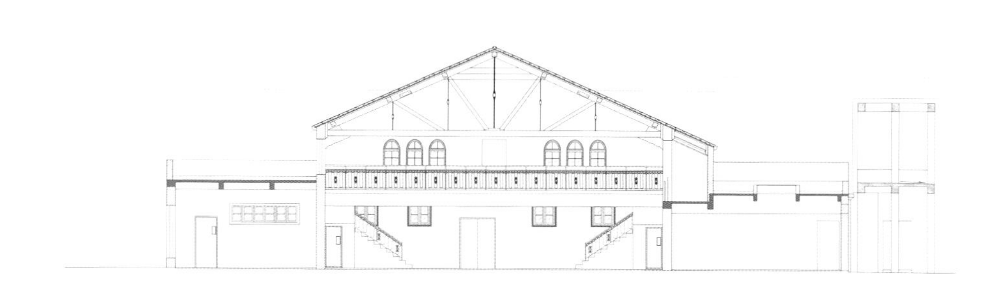

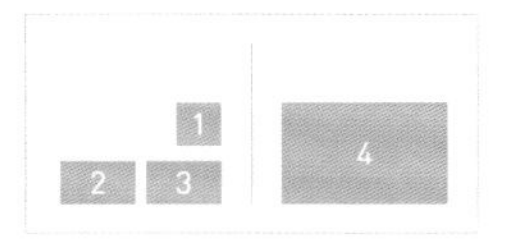

1. 旧时的哥伦比亚俱乐部
2. 哥伦比亚俱乐部立面图
3. 哥伦比亚俱乐部剖面图
4. 改造后的哥伦比亚俱乐部轴测渲染

1. Columbia Club in the Early Years
2. Elevation of the Columbia Club
3. Section of the Columbia Club
4. Axonometric Render of the Columbia Club after Reconstruction

“建筑设计是一场混沌的冒险，而‘大’则是对抗整合现代都市片段化和混乱的主要方式之一。”

关于建筑师

库哈斯及其合伙人在 1975 年创办的 OMA 最为国人熟知的作品便是央视“大裤衩”。父亲是作家、爷爷是建筑师的库哈斯早年走的是文艺路线，24 岁才到伦敦 AA 建筑学院学习，后至康奈尔和纽约深造。库哈斯回 AA 教书时的学生扎哈 • 哈迪德短暂地加入过 OMA，一起赢得了一些竞赛，但相关项目都未获建。如果说 1978 年成书的《癫狂的纽约》奠定了库哈斯事业的基调，那其理念的首次实现就是 1987 年建成的荷兰舞蹈剧院。随后 OMA 迅速扩张，代表作有西雅图中央图书馆等。库哈斯在哈佛教书时带领刘宇扬等人研究中国，并在 OMA 成立了关注非建筑实践的 AMO 设计研究工作室。在标新立异之外，OMA 近年也致力于城市更新和历史建筑改造，接手的最古老的“文物”是始建于 1228 年的威尼斯德国商馆，而由 OMA 合伙人克里斯 • 范 • 杜恩（Chris van Duijn）主持规划的“上生新所：哥伦比亚圈”则试图把上世纪初上海公共租界内侨民聚居之地重塑为当今沪上著名的公共空间。

ABOUT THE ARCHITECT

The most well-known work of OMA, founded in 1975 by Koolhaas and his partners, is the CCTV's "Big Pants" building. With regards to the background of Koolhaas, his father was a writer and his grandfather was an architect. As such, in his early years Koolhaas veered towards the literary and artistic route. At the age of 24, he went to AA Architecture College in London and then to Cornell University in New York for further study. During the period that Koolhaas returned to AA to teach, where Zaha Hadid, a student at the time, briefly joined OMA and together the partnership garnered prizes in a variety of competitions. Despite this, none of the award-winning projects were built. If *Dilurious New York*, which was published in 1978, laid the tone of Koolhaas's career, then its first realization was the Dutch Dance Theatre, which was built in 1987. Subsequently, OMA expanded rapidly, and its representative works included the Seattle Central Library in addition to myriad other projects. During his time as a professor at Harvard, Koolhaas led Liu Yuyang and others to study China in greater detail, and established the AMO Design Research Laboratory focusing on non-architectural practices in OMA. In addition to innovations, OMA has devoted itself to urban renewal and the renovation of historic buildings in recent years. The oldest "cultural relic" it has inherited is the Venice German Chamber of Commerce, which was built in 1228. The Historic Columbia Circle Regeneration, which was planned by Chris van Duijn, a partner of OMA, is an attempt to reshape the settlements of overseas Chinese in Shanghai's public concessions from the beginning of the last century into a famous public space in modern-day Shanghai.

雷姆 • 库哈斯
Rem Koolhaas

1.CCTV
2. 红毯广场景观示意图
3. 重建后的游泳池
4.1930 年代的哥伦比亚俱乐部
5. 重建后的哥伦比亚俱乐部游泳池
6.1945 年左右到的哥伦比亚游泳池与健身房
7. 改造后的哥伦比亚俱乐部

1. CCTV Building
2. Landscape Scheme of the Red Carpet Plaza
3. The Pool after Reconstruction
4. Columbia Club in 1930s
5. The Pool after Reconstruction
6. The Pool and the Gym in 1945s
7. Columbia Club after Reconstruction

23 Environment Upgrade of the Five Dragon Temple

五龙庙环境整改设计

古刹生息，一千年的两个故事

建筑设计：都市实践
项目地点：中国山西省运城市芮城县
完成时间：2016 年
建筑面积：267 ㎡
功能：环境整改，总体规划

Archi. design: URBANUS
Location: Ruicheng County, Yuncheng City, Shanxi Province
Completion time: 2016
Total area: 267m²
Program: Environment upgrade, Masterplan

位于山西省芮城县的五龙庙又名广仁王庙，是第五批全国重点文物保护单位，建于唐大和六年（公元 832 年），是现存最早的道教建筑。与五龙庙历史地位不相称的是其周边环境。庙前土坎下原有的五龙泉，因近年水位的下降已干涸。龙王庙祈雨文化的消失、乡村邻里中心的衰弱，都让五龙泉这一村民的精神中心沦落。

2015 年，万科锁定投资五龙庙环境整治工程，开启了“龙 • 计划”，把集资目标推向全社会，引起人们对古建筑遗产更多的关注。这是一次国家专项资金与社会资金合作进行文物保护事业的新尝试，也是在互联网平台上推广文保工作的新尝试，更是嫁接在世博会的国际平台上宣传中国文物和文保的新尝试，使这个千年古庙的文物本体在获得国家文保资金修葺之后，又获得了环境品质的改善，将一个孤立古庙转换为一座关于中国古代建筑的博物馆，融入当下生活。

环境整治设计围绕着两条线索展开。明线是以五龙庙为主体，展开一系列有层次的空间序列，并植入相关展陈，从而使观者能够更好地欣赏、阅读、理解文物；暗线则是通过提升五龙庙的环境品质和重新解读五龙庙，加强了这一场所的凝聚力，使村民重新聚集、交往在这一世代相传的公共空间，为当下农村精神价值的重塑创造出契机。

Situated in Ruicheng County, Shanxi Province, the Five Dragons Temple (Guang Ren Wang Temple) is listed as a class A cultural relic by the National Cultural Heritage Conservation Bureau in China. Built in 832 A.D. during the Tang Dynasty, it is the oldest surviving Taoist temple. Sitting on a raised ridge above its surrounding village, the temple itself is segregated from the everyday lives of the villagers. The original picturesque view of the temple has also lost its charms due to the increasing exacerbation of the environment. Furthermore, modern irrigation techniques has replaced the rite for praying rain, and thus turning the Five Dragons Temple from a spiritual centre to rubbish.

In 2015, Vanke Group initiated the "Long Plan" to raise fund to revitalise the environment of the Five Dragons Temple. This plan also helps to arise the public awareness on this historical preservation project. This initiative would then go on to become the first time where the government and private funds cooperated for cultural relics preservation, as well as the promotion of cultural protection through the platforms of internet and the international Expo.

The design of the environment uplift for the Five Dragons Temple is centered around two themes. An outstanding theme is to create layers of overlapping spaces around the main building to tell the story of the temple history and ancient Chinese architecture. Through this theme, people would learn about the knowledge of traditional Chinese architecture to better understand the importance for the preservation of heritage. The latent theme is to restore the temple into an area of public gathering in the village, and to give an alleviated environment to encourage contemporary lifestyles in coherent with the realms of ancient architecture.

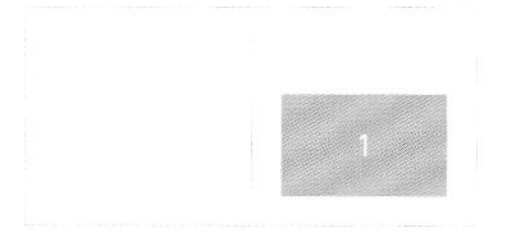

1. 五龙庙轴测图

1. Axonometric Drawing of Five Dragon Temple

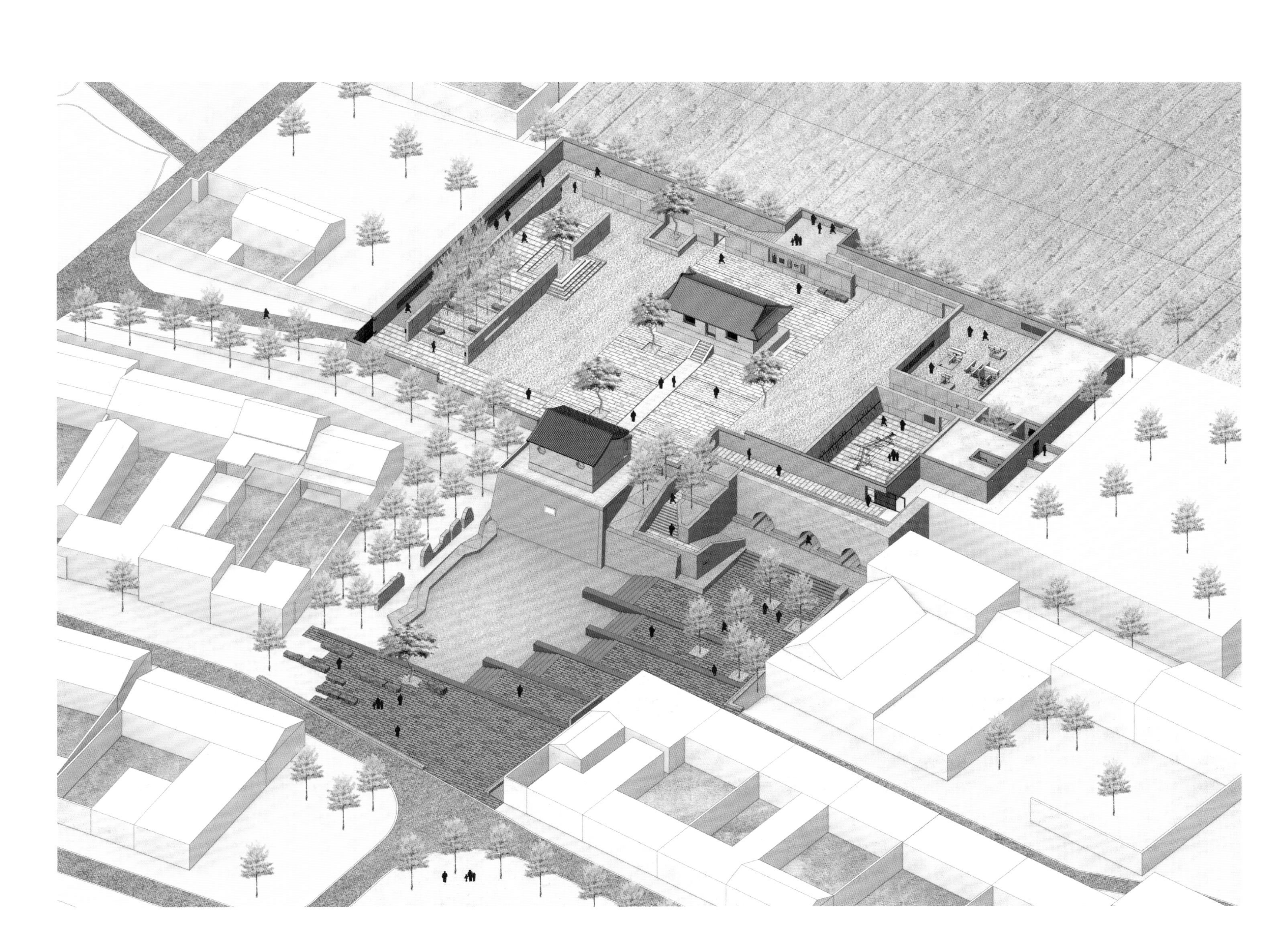

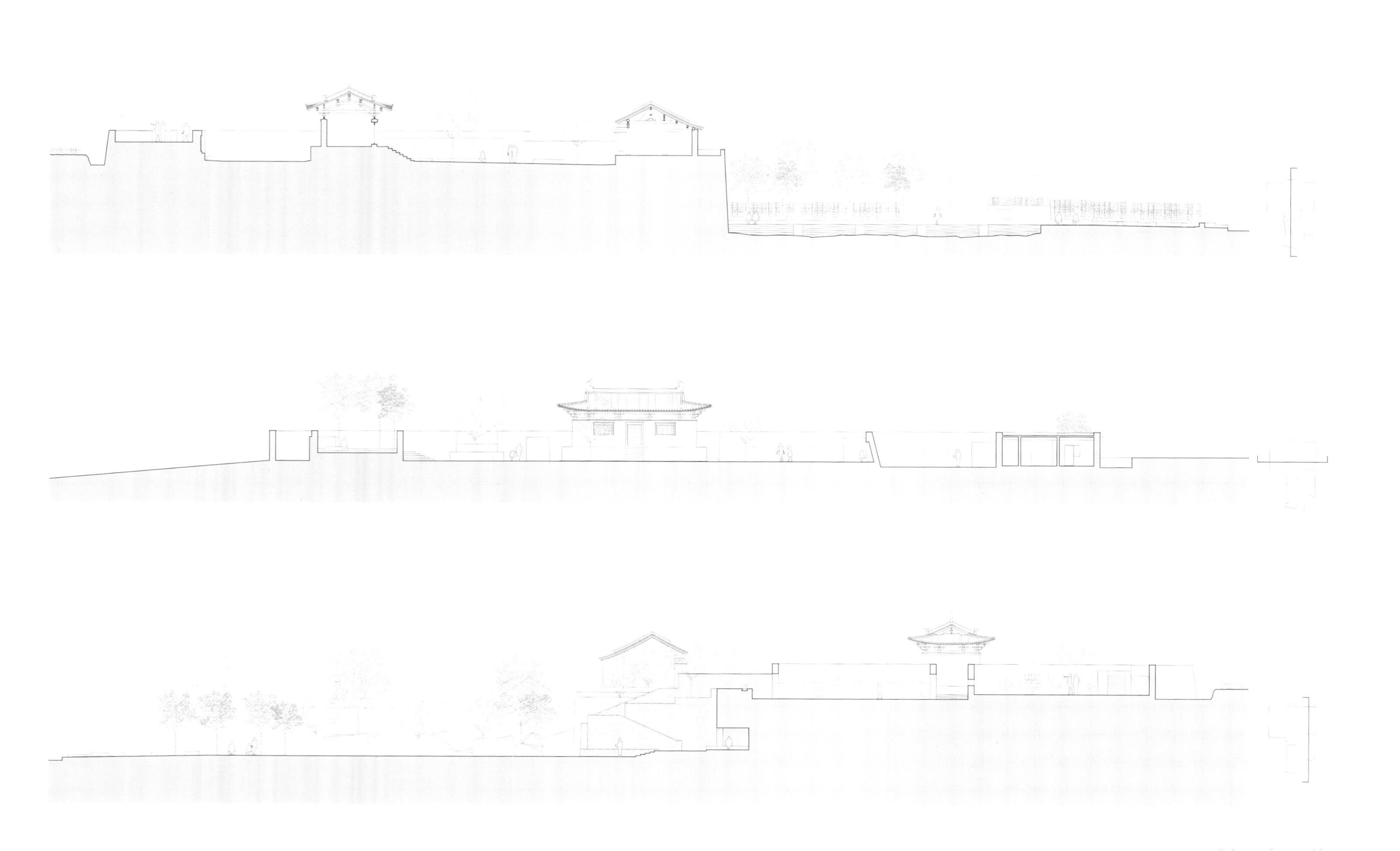

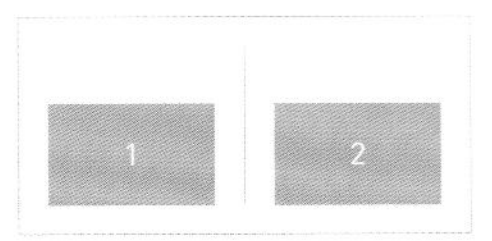

1. 剖面图
2. 改造后的五龙庙环境实景

1. Sections
2. Five Dragon Temple after Environment Upgrade

“每一次当我们将一个破旧的老建筑经过设计改造而再生一个脱胎换骨的新建筑生命，我们都可以得到一个喜悦的回报和延续历史的慰藉，并引导人们创造他们所向往的生活。”

关于建筑师

王辉在都市实践的三位创始人中最年轻，也最有书生气质，做事细腻认真，擅长旧建筑改造。他从 2003 年起主持北京分公司的运营，参与京津唐和长三角等地区许多城市和建筑的设计工作，其中的唐山城市展览馆及公园最为知名，而近年主持的山西芮城五龙庙环境整治则引发业内对文保与改造的热议。

唐山城市展览馆及公园原为唐山面粉厂，保留了四栋日伪时期的弹药库和两栋地震后建的粮仓作为展示厅，既纪念历史与重生，又与周边环境形成一片开放宜人的公共空间。而五龙庙是中国屈指可数的唐代建筑遗存，也是现存最早的道教建筑，但在改造前却沦为村里的垃圾场，与周围的新房格格不入，这促使王辉筑起立墙，区隔空间，重构本该属于那片土地的气场。对他来说，他不是在发明场所，而是在找回中国庙宇的空间节奏，恢复人们尊重传统的心灵秩序。

ABOUT THE ARCHITECT

Out of the three founders of URBANUS, Wang Hui is the youngest one with a scholar temperament. He is good at revitalizing old architecture with a fine and exquisite manner. Wang Hui is the associate of URBANUS Beijing office since 2003, and had participated in many urban planning and architecture design projects in Beijing-Tianjin-Tanshan and Pearl River Delta areas, famous outcomes include Tangshan Urban Planning Museum and Park, and his latest hosting project for the environmental upgrade of the Five Dragons Temple (Ruicheng County, Shanxi Province) had raised heated debate for historical site preservation and reconstruction in the architecture realm.

王辉
Wang Hui

Tangshan Urban Planning Museum and Park was reconstructed from the old flour mill, the original four warehouses built during the Second World War and another two built after the earthquake are all preserved as showrooms in a museum park, keeping the memory of the old city while mingling with the surrounding environment as an open public area. The Five Dragons Temple is listed as a Class A cultural relic by the National Cultural Heritage Conservation Bureau in China and is the oldest surviving Taoist temple built in Tang Dynasty. The Temple, before the upgrading process, had lost its charms and segregated from new development housing, then become a local dumpsite. Wang Hui had built layers of wall to overlap the spaces, rejuvenated the spirit of the old temple. He is never "inventing" a space, but rater to obtain the original spacial rhythm of the Chinese temples, and encourage a respectful mentality towards tradition and heritage preservation.

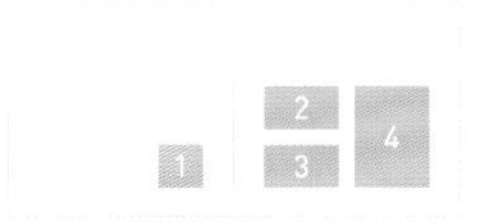

1. 唐山城市展览馆及公园
2. 龙泉遗址
3. 斗拱庭
4. 从入口夹道看五龙庙

1. Tangshan Urban Planning Museum and Park
2. Heritage Site of Longquan
3. Courtyard of Brackets
4. View of Five Dragon Temple from the Entry Lane

PART H Insta-Famous Architecture
网红建筑学

24 YOUXIONG Apartment

“有熊”

贝家有清祉，岁月洗铅华

建筑设计：B.L.U.E 建筑事务所

项目地点：苏州姑苏区敬文里 29 号

设计时间：2017 年 1 月—2017 年 5 月

完成时间：2017 年 5 月—2017 年 9 月

建筑总面积：2,500 ㎡

功能：公寓

Arch. design: B.L.U.E Architecture Studio

Location: No. 29 Jingwenli, Gusu District, Suzhou

Design process: 2017.01—2017.05

Completion time: 2017.05—2017.09

Total area: 2,500m²

Program: Apartment

项目位于苏州老城区的一处古宅，宅院占地面积 2500 平方米，始建于清代，前后共四进，其中四栋建筑是清代的木结构古建筑，另四栋为后来扩建的砖混结构建筑。设计内容包括古建筑和现代建筑改造，室内设计及庭院改造，将老宅院变身为现代文旅公寓。

设计基本沿用了原有的庭院布局。对于清代古建改造部分，设计保留了全部的木结构，并在内部增加了空调和供暖系统，以及卫生间、淋浴间等现代生活所必需的功能。外立面改造去除原有木结构表面的暗红色油漆，改为传统大漆工艺做的黑色，与原木色门窗结合，展现出老宅古朴素雅的气质。室内材质的选择方面，采用黑胡桃木材、天然石材等自然材质，忠实于材料本身真实的质感，延续古朴的氛围。砖混建筑改造的部分，则去除了原先立面上的仿古符号，新做的黑色金属凸窗使用的是简洁而纯粹的现代语言。室内使用原木色家具，与古代建筑室内的深色黑胡桃形成对比，更具有轻松舒适的现代气息。新与旧有着各自清晰的逻辑，在对比和碰撞中和谐共存。

整个宅院在历史上是属于一户人家的私宅，虽然要改造成现代公寓，但设计理念是希望延续老宅原有的精神和空间体验感，而不是将宅院割裂成一个个孤立的客房。对于每个入住的客人，不仅有自己的私密空间，更能走出来在整个园子里与其他人交流。整个园子除了 15 个房间作为客房，另外超过一半的空间都作为公共空间利用，例如公共的厨房、书房、酒吧，甚至是公共泡池。做饭、健身、休闲娱乐等功能不但可以在自己的房间里完成，也可以在园中和他人一起以共享的模式实现，家的意义在概念和空间上都被扩大了。整体的功能布局在庭院从南侧入口向北侧层层递进的同时，完成公共向私密的过渡和转化。

庭院是苏州古宅中最美的空间，庭院成为另一个设计重点。老宅院里，每个古建筑都有一个独立的庭院，在设计中把原本格局中没有庭院的房间，也特意留出一部分空间作为庭院使用。住宅不再是封闭的，室内与室外相通，庭院与庭院相连，延续了苏州园林的情趣，空间随着人的行走变化流动，人的感官体验是动态的。其中的亮点是入口空间，原先的停车场被改造成了石子的庭院和水的庭院，穿过竹林肌理的现浇混凝土墙面，回家的客人从外面的城市节奏自然地转换到园林宁静自然的氛围里。水池中的下沉座椅，让人们在休息时更加亲近水面和树木，带来不一样的视角和体验。通过庭院的改造，动和静、城市和自然，达成了最大程度的和谐。

古宅的改造是一种与历史的对话，在城市人越来越倾向独居生活的个体时代中，希望通过苏州古宅的改造，创造一种打破私密界限，让人与人、人与自然都能产生交流的空间，这是一种对新的生活方式的探索，也是对于古城更新模式的一种新思考的开始。

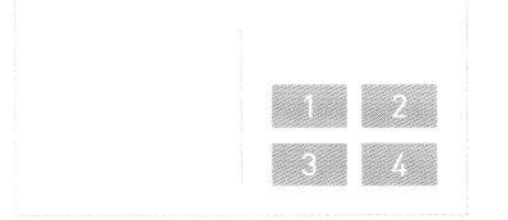

1. 总平面图
2—4. 平面图

1. Masterplan
2—4. Building Plans

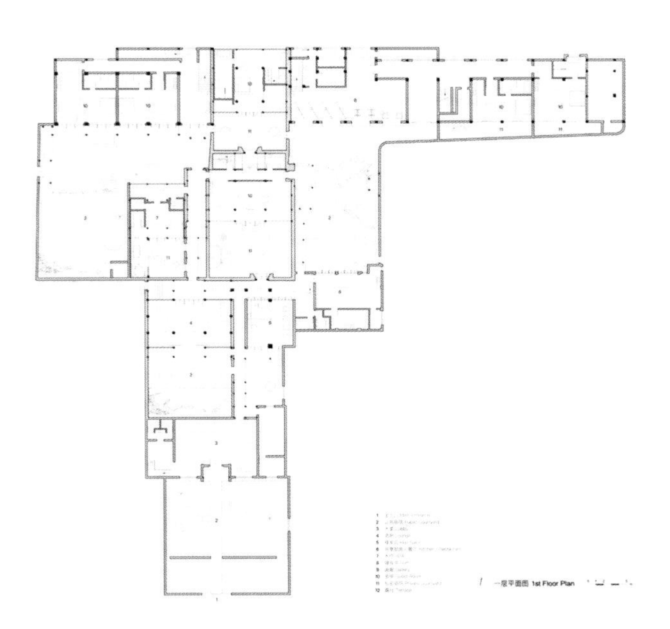

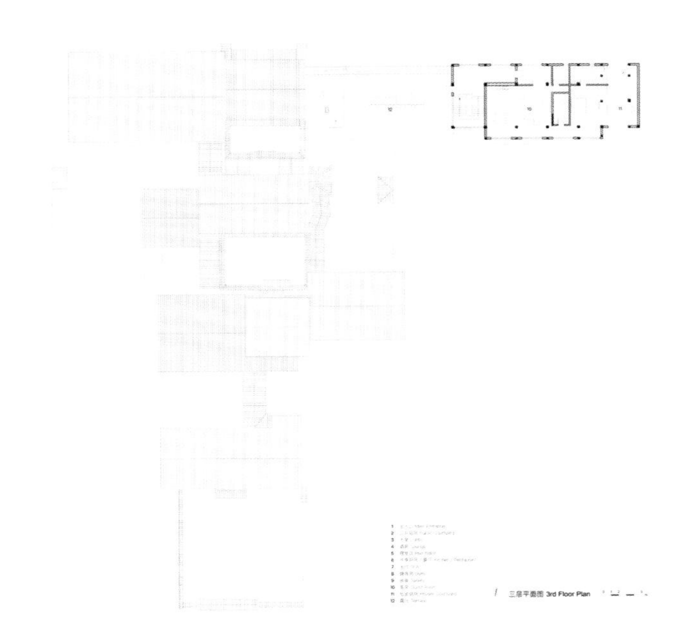

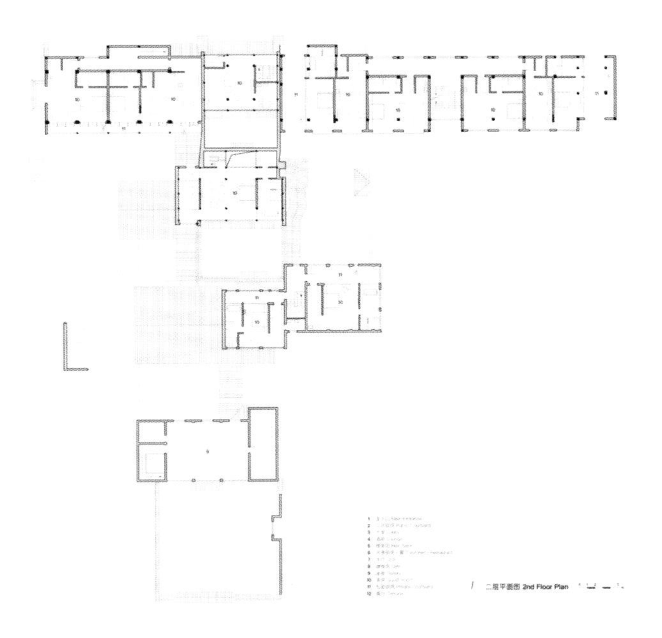

Located in the core of the historic town of Suzhou, China, the site covers about 2,500 sqm, once was the residence of family Bei. The traditional Chinese garden consists of four ancient wooden structure architectures with courtyards built in the Qing Dynasty, dating back over a hundred year of history, and other four buildings built in the 1990s of concrete structure. The renovation project aims to transform the historical house into a modern apartment.

For the renovation of the four old houses, all the original wooden structure is preserved with simple reinforcement and restoration. Since the lack of maintenance, the building status no longer suits modern lifestyles, the main focus is to resolve functions that can meet the needs of modern daily life, such as air conditioning, heating system, shower room etc. The red paint from the original wood structure is removed and changes to black paint using the traditional lacquer process, together with natural wood frame of the windows and doors, reflect the beauty and elegance of the old architecture. Materials with natural textures such as stones, walnut wood are used for the interior space, setting a simple and natural tone. For the renovation of the concrete structure buildings, the design goes for a more relaxed and modern atmosphere using pure and minimalist language. Comparing to the old houses, materials with lighter color are used such as oak wood and light grey terrazzo floor. By boldly introducing the modern design into the historic garden, the juxtaposition of old and new shows an interesting dialogue between the tradition and the modern lifestyle.

The main theme of the design is to inherit the spirit of the historic house not just by preserving and restoring its appearance but most importantly by recreating its spatial experience. Even though the house is to be transformed into a modern apartment, we do not want it to be separated into several isolated rooms, in fact, more than half of the space is used as public space, such as shared kitchen, shared study room, art galley, public bars and spa... Apart from the private room, the guests will also have the whole garden as the extensions of their home where they can communicate with the others. The definition of home have been expanded both in terms of concept and space.

Courtyards are considered to be the most beautiful spaces in traditional Suzhou houses, hence the courtyard design become the vital element of the restoration project. In the old housing, each building would have its own individual courtyard, for this particular design additional spaces were dedicated to those rooms that originally had no courtyards for a more open and adjoining experience. Through the layers of courtyards and corridors, the spaces are flown through the walking experience. The old parking lots are transformed into a courtyard with pebbles and water surfaces, as a key entrance space, the guests would transit from the bustling urban scene into a tranquil natural space along a concrete wall with bamboo texture surface. The seatings are located near the pond which makes an intimate connection between the water surfaces and the woods, merging the different elements between city life and private nature scene seamlessly.

Restoring and refurbishing the ancient house is like a conversation between now and history. Through the openness of the design, the sealed-up individual boundaries are blurred, enabling the connection between human and nature, hence between people. This is not only a new attempt to a lifestyle but also new thoughts on overall old-town upgrading system.

1. 轴测图

1. Axonometric Drawing

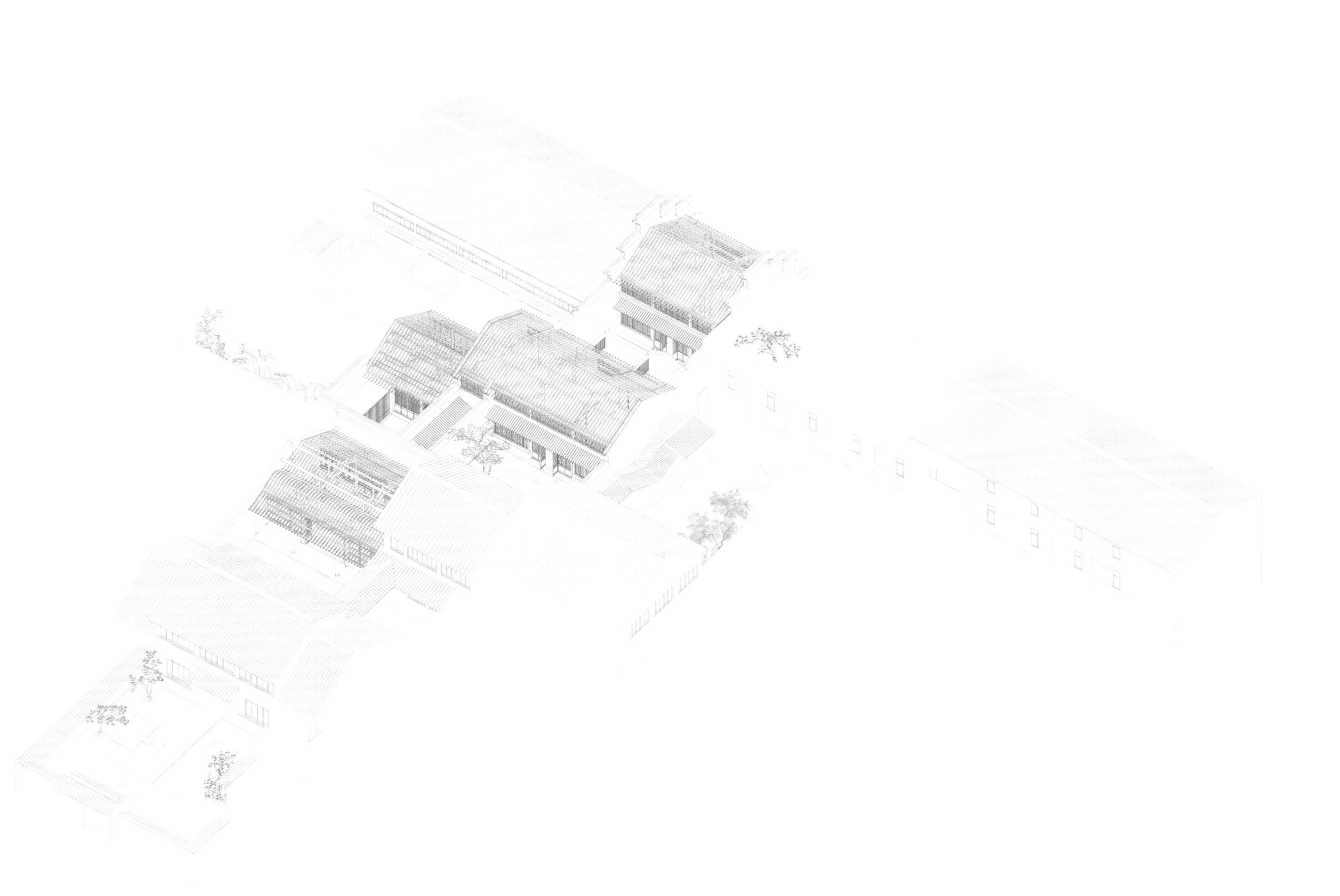

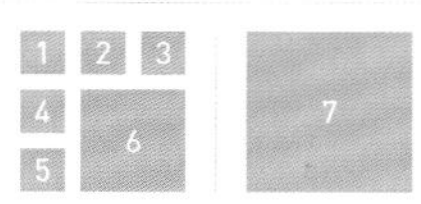

1—7. 实景照片

1—7. Photos

"设计是解决社会问题的工具，最纯粹的设计便是与大自然相融，在自然而然中到达期待的目的地。"

关于建筑师

青山周平在真人秀《梦想改造家》中爆改北京胡同老房而走红，是粉丝心中当之无愧的"男神"。毕业于大阪和东京大学的他，对奥运前中国的变化感兴趣而到朋友在北京的事务所实习，却一下在那里工作了七年，设计了天津塘沽实验学校远洋城校区，直到 2012 年受聘北工大，并于 2014 年与太太藤井洋子共同创建 B.L.U.E. 建筑设计事务所。尽管父亲就是建筑师，但青山只把做建筑看作一种探索现代生活、改善社会环境的手段，在清华继续读博，还在南锣鼓巷的胡同里一住就是十年，因为有很多共享空间的市井生活才更本真，比私密封闭更接近他对"家"的理解。他对老房子和小户型清爽明快又极具人情味的改造让吃瓜群众大开眼界，但他近来将苏州贝聿铭家族的老宅嘉园变成日系风格的有熊公寓却引发争议。不过青山说的没错："有时候很有价值的创新是从外面来的，因为外国设计师没有那么多限制，所以会创造新的传统，这个可以说是文化上的自由。"

ABOUT THE ARCHITECT

Shuhei Aoyama became popular from the reality show "Dream Home" in which he renovated old Beijing Hutong homes. He is a well-regarded "male idol" in the hearts of his fans. Having graduated from Osaka and Tokyo University, he found himself fascinated with the changes in China before the Olympics and interned together with his friends in Beijing's office. He worked there for seven years, during which he designed the Ocean City Campus of Tianjin Tanggu Experimental School, until in 2012 when he was hired by the Beijing University of Technology. In 2014, he co-founded B.L.U.E. Architectural Design with his wife, Yoko Fujii. Although his father is also an architect, Shuhei only regards architecture as a means of exploring modern life and improving the social environment. He has continued to study at Tsinghua University and live in the alley of Nangulou Lane for ten years. Because there are many shared spaces in the city, the life is more authentic and closer to his understanding of "home" than private closures. His refreshing, lively and humane renovation of old houses and small apartments has opened the eyes of the audience, but his recent transformation of Jiayuan, an old house of the I.M. Pei family in Suzhou, into YOUXIONG Apartment in the Japanese style has aroused controversy. However, Shuhei's words ring true: "Sometimes, the most valuable creations are coming from the outside, since the foreign architects do not have the restrictions, and they could create new traditions, thus we call it the freedom in culture."

青山周平
Shuhei Aoyama

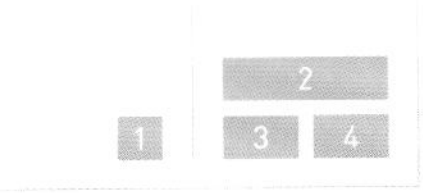

1.DOE 门店
2—4. 客房效果图

1. DOE Storefront Design
2—4. Room Renders

25
Stage of Forest

吉林松花湖滑雪场
森之舞台

建筑设计：META 工作室

项目地点：中国吉林省松花湖度假区

完成时间：2017 年

建筑总面积：277 ㎡

功能：观景平台，公共活动空间

Arch. design: META-Project

Location: Songhua Lake, Jilin Pr ovince, China

Completion time: 2017

Total area: 277m²

Program: Viewing platform, Public activity area

森之舞台，位于吉林市松花湖风景区。它坐落在大青山顶，森林边缘的山坡之上。

夏季，这里被浓郁的绿荫环绕；到了冬季，粉雪覆盖一切，形成了一条绝佳的野雪道。这一“地景标志建筑”的选址，基于对场地条件的全方位考量。建筑师希望尽量减少建筑对现有植被的影响，同时保证观景平台上视野的最佳角度，由此确定了基地位置和三角形舞台的基本形状。

森之舞台就从起伏的地景中缓缓升起，如漂浮在水面的一片树叶，悬挑于雪道之上。三角形平台的两条短边，按两条访客路线设置：自林间小路走来，或从雪道滑来。如此，舞台非但没有阻碍人们望向湖面和山峦的视线，更会给行人一种建筑形体与周围山景间不断变动的视觉张力。由于上部的木质舞台和下部的混凝土基座朝向景观不同，两个体量错位相接，建筑最终呈现出一种扭转的姿态。

建筑将粗粝的材料物质性和富于感观体验的空间形态融于一体。通过对建筑体的扭转、翘曲、直线与弧线墙面不断转换等方式，塑造着观者细微的感受差异。从山顶一路走来，“森之舞台”如同一块地景中自由飘起的玛尼石板，悬浮在如竖立的卵石般的混凝土“基座”之上。随着观者逐渐走近，先前感觉深沉的建筑体，在阳光下烧杉板的表皮隐约泛出银色的光泽；最终抵达建筑时，烧杉板皲裂的表面、清水混凝土表层的木纹理变得清晰而几可触摸。

建筑的内部组织更是一场精心策划的空间体验。进入延长的混凝土门斗后，你的眼睛会逐渐适应昏暗的光线，同时被雪道向远处延伸的景观所吸引。随后，狭窄的木楼梯指引着一路往上的唯一路径。到达观景层后一转身，刹那间，松花湖的湖景和蜿蜒山林横陈眼前，呈现出大自然令人窒息的美。气候无常，湖面时而清晰可见，时而雾气蒸腾；冬天的山林，偶尔还会出现“雾凇”奇观。

建筑师在平台的三角形体量中，斜切出一对椭圆形的洞口。一个洞口在屋面，等待雪花和阳光洒入室内空间。另一个洞口在地板，诱发舞台上下的人互动。平台内部整体为未做处理的红雪松木板，保留着原木色差，与外侧深色的烧杉板表皮形成色彩与质感上的强烈对比。

META 工作室认为，在自然中设计，就是搭建人与自然之间一个启发性的媒介。森之舞台不仅是一个观景平台，也是一个可以灵活用于活动、聚会、展览和工作坊的公共空间。它希望激发人们更多地探索自然，发掘人与自然的关系。而森之舞台本身，也将成为自然的一部分。

1. 山顶总平面图
2. 剖面
3. 观景台立面

1. Mountain Top Master Plan
2. Section
3. Elevatiovn of the Viewing Platform

Stage of Forest located at Songhua Lake Resort in Ji Lin,which is situated on a hillside between the forest and the slope.

The site is surrounded by luscious greenery in summer and covered by an overwhelming white snow in winter. As is a delicate site for a "Land(scape) Mark", one whose indefinite programming demands a careful degree of deliberation. The location and triangle shape of the "stage" was only determined after precise examination and deduction of the site condition, to minimize the impact for the existing vegetation and to maximize the view on the platform. While sitting on the hill, it is facing the Songhua Lake at a distance, who is famous for the rime in its surrounding areas.

The location and triangle shape of the "stage" was only determined after precise examination and deduction of the site condition, as one descends from the mountain top,the "Stage" rises slowly above the undulating landscape.The building results in a twisted gesture between the wood "stage" and the concrete "base" .

As one descends from the mountain top, the "Stage" rises slowly above the undulating landscape, in a way like a piece of leaf floating on the water. Positioned with the 2 side-line along the approaching eyesight of the visitors from two routes: the trail in the woods and the ski-slope. Not only it doesn't obstruct the view to the lake and mountains, it even enhances the experience by inducing ever-changing tension between the cantilever and the surrounding landscape. The entire "stage" is like growing out from the mongolian oak forest, and cantilevering on top of the ski-slope. Because the orientation of the distant view (horizontal unfolding lake), and the close view (vertical extending slope) is at a different angle, the building results in a twisted gesture between the wood "stage" and the concrete "base".

The interior is choreographed through a carefully plotted experience. Upon entering the concrete vestibule, in the moment your eye adjusts to the dimmed light, a vertical view along the stretching slope will catch you first, then a narrow staircase hints the only way of elevating. When you arrive at the platform level and turn around, what suddenly opens up to you is a great panoramic view of the Songhua Lake, winding in-between the hills, clear or hazed by with the ever-changing mist, an exceptional vista that is breathtakingly beautiful and magical.

A pair of ovals openings cut through the volume, one on the roof leading sunlight and snowflake into the space, the other one on the floor intriguing interaction between people above and under the "stage". The red cedar wall has been left untreated and is vivid in color shades, in contrast to the building's dark charred-wood-shingle (Shou-Sugi-Ban) exterior.

META-Project believes, designing in nature is to introduce an enlightening medium between nature and people. The "Stage of Forest" is not just a look-out, it is a flexible public space that can hold events, exhibitions, or workshops... The building is intended to stimulate people to come up with more ideas of exploring their relationship with the nature, and itself also becomes part of the nature.

“面对当下新的挑战，建筑必须从一种被不断累积的种种‘事物’所模糊了的状态中解脱出来，而达到复杂性和清晰性的新高度。”

关于建筑师

王硕从小在北京大院长大，清华毕业后去了美国莱斯大学，却一直着迷于亚洲城市的原生活力。他在莱斯写了硕士学位论文《狂野北京》(*Wild Beijing*)，努力理解建筑与城市文化生活的关联。随后他到鹿特丹 OMA（大都会建筑事务所）从事城市更新和规划，但为了做“对中国城市有意义有贡献的项目”，王硕与合伙人在纽约创立 META 工作室并于 2009 年到北京设立办公室。META 一半做设计，一半做城市研究，想让城市重新成为促进人们交流的日常空间。其研究和实践方向包括“新型城市文化的激活装置”、“混合型的社区”以及“青年人户内生活方式的转变”，而森之舞台就是继葫芦岛海滨文化展示中心、水塔展廊之后 META 在东北建成的第三个小而美的“激活装置”，既是人与自然间极具张力和启发性的媒介，又在人与城市间巧妙地串起一系列跟山地相关、超越日常的活动事件，发掘了建筑之于社会与文化的丰富潜能，重构了设计之于社会与环境的有益作用。

ABOUT THE ARCHITECT

Grew up in native Beijing courtyard, despite his overseas study experience in Rice University in the US, Wang Shuo had always obsessed with the raw energy in Asian cities. His master thesis "Wild Be[ij]ing" had shown great efforts in understanding relations between architecture and urban cultural life. Wang Shuo had previously worked at OMA in Rotterdam in urban revitalization and planning, to pursuit his dream of "realizing meaningful projects for Chinese cities", he then co-founded META in New York and set up his Beijing office in 2009. META emphasized on both designing and urban research, seeking a better cityscape with encouragement of daily interactive activities. After research topics such as "Activating devices for new urban culture", "Hybrid Community" and "Indoor lifestyle transformation of the youth"; The Forest Stage project was the third mini Activating Device META had produced in Dongbei Province after Huludao Beach Exhibit Center and Water Tower Pavilion. The Stage, acts both as an inspirational medium between human and nature, also as a mountain related event-stimulator between city and the site. The Forest Stage project had explored great potential in the value of architecture to society and culture, and restated the benefits of design to the social environment.

王硕
Wang Shuo

1. META 东北项目
2. 森之舞台夜景

1. META Projects in Northeast China
2. Nightview of the Forest Stage

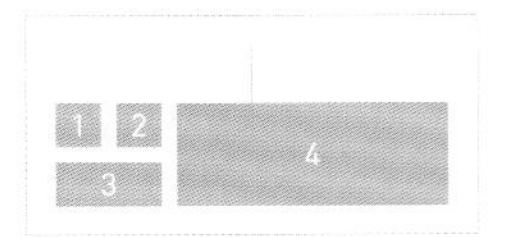

1. 仰望森之舞台悬挑处
2. 舞台室内示意
3. 冬林中的森之舞台
4. 全景图

1. Looking up the Cantilever of the Stage
2. Interior of the Stage
3. Forest Stage in Winter Woods
4. Panorama View of the Stage

衷心感谢以下各方对本书出版的大力支持

万科企业股份有限公司

北京国际设计周 – 北京科意文创企业管理有限公司

本书收录项目的各建筑师事务所及建筑师

同时也感谢

苏州市及姑苏区人民政府

苏州国家历史文化名城保护区管理委员会

对《大象无形 FOOTPRINTS》建筑展的大力支持

This book has been made possible with the generous support of the follow organizations

China Vanke Co., Ltd

Beijing Design Week - Keyi China Design Investment

Participating architecture firms and architects of this book *Footprints*

We would also like to thank the following organiazations to host the architectural exhibition of *Footprints*

Suzhou and Gusu District Municipal People's Government

Suzhou Bureau of the National Historic Preservation Commission

FOOTPRINTS 25 COLLABORATIVE WORKS OF GLOBAL ARCHITECTS AND VANKE REGARDING CHINA'S RURAL AND URBAN DEVELOPMENTS

大象无形 中国城乡建设的探索和引领 — 25 个世界建筑师与万科的合作与实践

出 品 人：韦业宁

策　　划：李长乐 / 金韵韵

策划助理：吴江源

编　　辑：吴文一 / Wang Yile / 蔡雨钱

编辑助理：杨尚龙 / 王潇崡

统　　筹：朱天恩 / 朱小风

撰　　稿：蔡雨钱

翻　　译：Wang Yile / Daniel Lenk

设　　计：沈康

插　　图：崔蕾

图书在版编目（CIP）数据

大象无形 / 吴文一主编 . — 北京 : 东方出版社 ,
2019.3
ISBN 978-7-5207-0748-0

Ⅰ . ①大… Ⅱ . ①吴… Ⅲ . ①建筑设计—作品集—世
界—现代 Ⅳ . ① TU206

中国版本图书馆 CIP 数据核字 (2019) 第 008182 号

大象无形
(DAXIANG WUXING)

主　　编：吴文一
出版统筹：吴玉萍
责任编辑：罗佐欧
出　　版：东方出版社
发　　行：人民东方出版社传媒有限公司
地　　址：北京市朝阳区西坝河北里 51 号楼
邮政编码：100028
印　　刷：鑫艺佳利（天津）印刷有限公司
版　　次：2019 年 3 月第 1 版
印　　次：2019 年 3 月第 1 次印刷
开　　本：889 毫米 ×1194 毫米 1/12
印　　张：20.5
字　　数：250 千字
书　　号：ISBN 978-7-5207-0748-0
定　　价：298.00 元
发行电话：（010）85924663 85924644 85924641
